MAN, ENVIRONMENT AND DISEASE IN BRITAIN

Man, Environment and Disease in Britain

A Medical Geography of Britain
through the ages

G. Melvyn Howe, MSc, PhD
Professor of Geography
in the
University of Strathclyde

BARNES & NOBLE BOOKS : NEW YORK
(a division of Harper & Row Publishers, Inc.)
DAVID & CHARLES : NEWTON ABBOT

Published in the U.S.A. 1972 by
HARPER & ROW PUBLISHERS, INC.
BARNES & NOBLE IMPORT DIVISION

ISBN 0 7153 5547 3 (Great Britain)
ISBN 06 4930203 (United States)

Set in 11 on 13pt Garamond
and printed in Great Britain by
Richard Clay (The Chaucer Press) Ltd., Bungay, Suffolk
for David & Charles (Publishers) Limited
South Devon House, Newton Abbot, Devon

ON AIRS, WATERS, AND PLACES

1. Whoever wishes to investigate medicine properly, should proceed thus: in the first place to consider the seasons of the year, and what effects each of them produces . . . Then the winds, the hot and the cold, especially such as are common to all countries, and then such as are peculiar to each locality. We must also consider the qualities. In the same manner, when one comes into a city to which he is a stranger, he ought to consider its situation, how it lies as to the winds and the rising of the sun; for its influence is not the same whether it lies to the north or the south, to the rising or to the setting sun. These things one ought to consider most attentively, and concerning the waters which the inhabitants use, whether they be marshy and soft, or hard, and running from elevated and rocky situations, and then if saltish and unfit for cooking; and the ground, whether it be naked and deficient in water, or wooded and well watered, and whether it lies in a hollow, confined situation, or is elevated and cold; and the mode in which the inhabitants live, and what are their pursuits, whether they are fond of drinking and eating to excess, and given to indolence, or are fond of exercise and labour, and not given to excess in eating and drinking.

The Genuine Works of Hippocrates
Francis Adams, 1849, Vol I, p 190

Contents

List of Illustrations

PLATES

FIGURES AND MAPS

TABLES

Preface

Over the last twelve or so years I have been examining the geographical distribution within present-day Britain of mortality from such 'degenerative' diseases as ischaemic heart disease, cerebro-vascular disease, chronic bronchitis, and the various cancers. The aetiology of these diseases is, to a large extent, still unknown but the mortality from each displays a distinctive areal pattern suggestive of environmental relationships. Prompted by this I decided to look at the killing and other diseases of earlier centuries from the geographical viewpoint to see if they reflected reactions by man to adverse factors in the environments of the day. These were retrospective studies and concerned infectious diseases for which the aetiology is known. They afforded the opportunity however of highlighting the effect of environmental circumstances on the micro-organisms which cause infectious disease.

The present book contains the results of these several studies, which ranged from pre-Norman times to the present. Infectious diseases and degenerative diseases are examined. The former constituted one of the fundamental facts in the lives of most people until almost within living memory, the latter have taken their place now as the principal cause of suffering and death in twentieth-century Britain.

It is likely that the task undertaken was too ambitious for one person and, in so far as it has involved many academic disciplines, mistakes are inevitable. It is to be hoped, however, that critics will not take me too much to task over minutiae and lose sight of the main purpose of the book which is to provide the intelligent layman with an historical survey of the intricate relationships of human disease and environment in Britain from an areal or geographical viewpoint.

No one can assess better than I the number of predecessors and contemporaries to whom I am indebted for this volume. The late Lord Nathan and the late Sir Dudley Stamp provided me with the initial inspiration. The help and kindness of many friends and acquaintances has since sustained my interest. The list of references at the end of the book includes but a fraction of the literature to which I owe the facts here presented. I should like to emphasise this and to express my gratitude to all who for reasons of practicability cannot be acknowledged individually. My sincere thanks are extended to all those to whom I am indebted.

My wife Patricia has been a constant source of encouragement. I acknowledge her forbearance with gratitude.

G.M.H.

Glasgow, November 1971

I
Introduction

The World Health Organisation, in its Constitution,[1] defines health as 'a state of complete physical, mental and social well-being, and not merely the absence of disease or infirmity'. Health is not an absolute quantity but a concept whose standards are continually changing in different lands with the acquisition of knowledge and the establishment of cultural objectives. To be truly healthy a person should enjoy a balanced relationship of the body and mind and complete adjustment to the environment. In this context 'environment' refers to the whole gamut of influences which impinge on man and affect his well-being: it includes man's physical surroundings of land, sea, and air, the viruses, bacteria, and other organisms of the biological environment together with the socio-economic complexities of the human environment. In contrast, sick or diseased persons represent maladjustment or maladaptation in an environment. Disease is a reaction for the worse between individual man and the stresses, strains, and other adverse factors of his surroundings, the response being conditioned by the genetic make-up (ie *internal* environment or inborn constitution) of the individual.

The present study examines the main diseases which, at one time or another throughout the ages, have affected the people of Britain. It is an interdisciplinary study, but since the emphasis is on the areal spread of disease, areal variability of disease, and areal relationships and interrelationships of environmental and aetiological factors, it is best considered as a study in medical geography rather than in epidemiology.

[1] Notes and References commence on p 245.

Medical geography may be defined as the comparative study of the incidence of disease and the distribution of physiological traits in people belonging to different communities throughout the world and the correlation of these data with features of the environment. Different ethnic, national, or social groups may be involved. These groups, may, however, live in close proximity or in different parts of the same country, as well as in countries remote from one another. The subject is closely related to epidemiology except that in medical geography the emphasis is on patterns of distribution and on the areal aspects of environmental relationships.

Much of the subject-matter of medical geography is as old as Hippocrates, and his *De Aëre, Aquis et Locis* (on Airs, Waters and Places) is well known. Even the term 'medical geography' has been current in Britain for eighty or so years, and was used by Dr Alfred Haviland in his *Geographical distribution of disease in Great Britain* (1892). Medical geography itself was somewhat usurped after Pasteur introduced the germ theory and heralded the science of bacteriology in the late nineteenth century. Now, however, with the increasing control or eradication of infectious diseases and, in so-called developed countries such as Britain, the ever-increasing concern with the degenerative diseases of later and middle life, the new methods and techniques of medical geography are being called upon to determine the geographical distribution of these diseases in an attempt to relate differences in incidence to local environmental factors and afford pointers to possible causal relationships.

Man is, and always has been, part of the ecosystem.[2] He is but one species in relation to the total evolutionary history. Even so, differing from other members of the animal kingdom which have either suffered extinction or undergone genetical evolution as a result of changes of environment, man has undergone cultural development which, in large measure, has overtaken his genetical evolution. Instead of the environment shaping his genetic destiny, man now shapes his environment. Of the changes he has invoked some, such as improved housing, water supply and sanitation, have been good, some, such as atmospheric and water pollution and excessive noise, are obviously harmful; others, such as nuclear explosives might well prove catastrophic. Nuclear radiations are immeasurably significant, not only in their capacity for environmental change but also for their effects on man's health and genetic stability.

Infectious diseases have been controlled in Britain chiefly by a combination of enlightened environmental changes, health legislation, education, and secondly by therapeutic advances. Health problems now are of a more chronic kind and include heart disease, cancer, bronchitis, and accidents. In large towns man lives in a polluted atmosphere to which his lungs respond with respiratory disorders and cancerous growths. The tempo of life is such as to create mental tensions and stress which, in their turn, encourage heart disease. Overcrowding in large communities assists the spread of droplet infections; motor vehicles constitute an ever-present accident hazard, and there is persistent and irritating noise. War itself has reached the annihilation stage, for it includes nuclear devices, germs, and chemicals.[3]

Yet technology helps modify and control the environment, medical care and research have taken the terror from many of the diseases that were once fatal and there are simple means of birth control for regulating the size of the population. There is no longer in Britain a poor, ignorant, and rural population exposed to hunger, malnutrition, and the killing diseases of infancy, childhood, and early adult life as in previous centuries. Instead the population is relatively affluent, informed and largely urbanised and industrialised. Nevertheless it is scourged by diseases of the cardiovascular system, respiratory system, by cancer, and mental illness.

Environmental improvements, particularly in housing, nutrition, and education, together with dramatic advances in medicine, have resulted in a far greater life expectancy for the average person. The people of Britain today are bigger and healthier than ever before. Expectation of life increased from about 18 years in prehistoric times to about 33 years in the Middle Ages and to near the biblical three score years and ten at the present day. In the 1860s expectation of life at birth for a boy was little more than

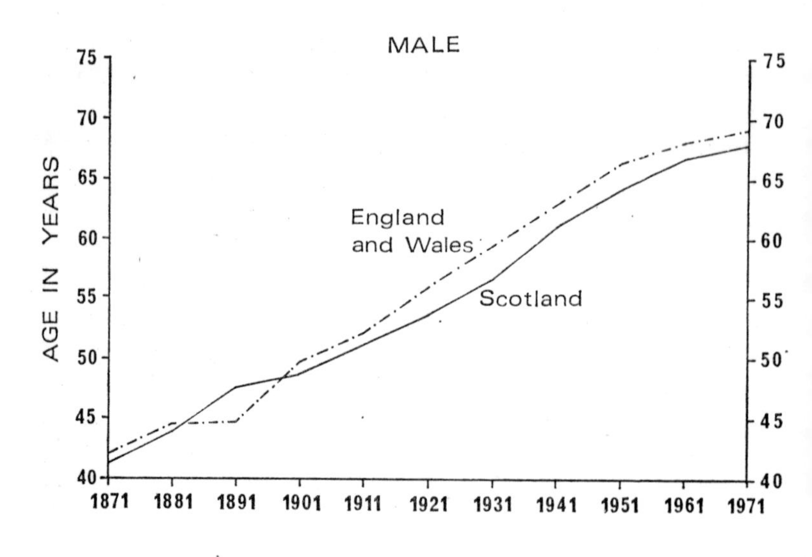

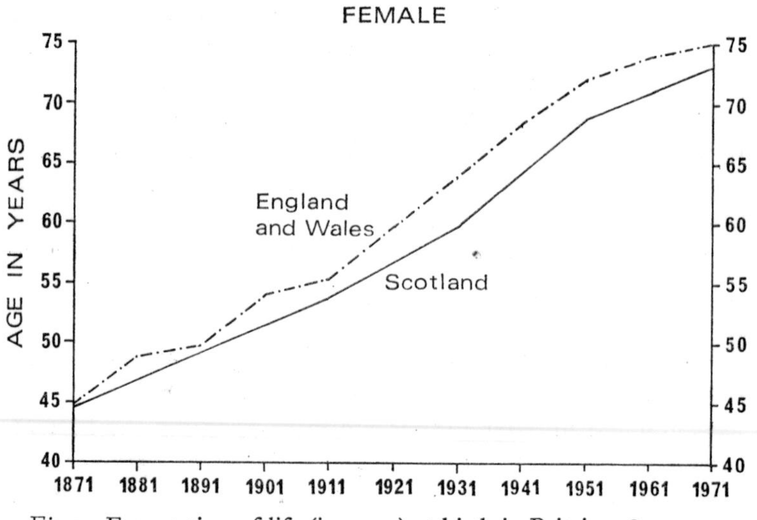

Fig 1 Expectation of life (in years) at birth in Britain, 1871–1971

43 years; in the 1960s it was about 68 years. In the early 1970s it is 69 years for men in England and Wales and 67½ years in Scotland. For women the expectation is 75 years and 73½ years respectively (Fig 1). 'Days of absence from work' or 'the number of visits to the general practitioner' probably represent more satisfactory indices of the general health of the community now than do death rates (see Chapters 13 and 14).

Disease is rarely the result of a single factor but rather of a combination of related factors. At a recent WHO conference on the prevention and control of cardiovascular diseases, a group of at least eight factors was deemed to be significant in coronary artery disease: high blood lipid levels (lipid being a general term that includes fats and fat-like compounds), hypertension, cigarette-smoking, physical inactivity, increase in weight, nervous stress, diabetes mellitus, and genetic factors.[4] Much speculation exists as to the role played by each of these factors and their relative importance. The obese are more prone to diabetes; people under stress take little exercise; racial and genetic differences are often associated with differences in diet. Obesity, exercise, and diet are themselves all interrelated. When examining the correlation between human diseases and individual environmental or genetic factors it is more or less inevitable that other factors will interfere with the conclusions. Causes of disease are rarely simple or even static; they are invariably multifactorial. Neither are diseases immutable throughout history. Modern descriptions of a disease need not necessarily conform to its course throughout history.

Since human response to environmental hazards is conditioned by the inborn constitution or genetic make-up of the individual, a brief outline of Britain's racial history is first presented (Chapter 2). This is followed by consideration of environmental hazards thought likely to promote disease in man (Chapters 3–5). In the remainder of the book (Chapters 6–14), a holistic approach is adopted. Attention in this, the larger part of the book, is directed to a demonstration of the synthesis of relationships and inter-relationships of people, environment, and disease at selected stages in British history. For each 'stage' or 'period-picture', a normality[5] or normal state is assumed, during which man and environment are considered to be in a state of symbiosis or mutual equilibrium. It might be questioned if such a state ever really existed and if it did, how it should be measured. A range of normal values for individual characters of structure or formation is accepted and defined by statistical constants, but man is something more than his parts. Assessments of a normal person vary from age to age, and also among clinicians, actuaries, sociologists, psychologists, and the various branches of the health service. In each case 'normality' implies fitness for a purpose. But for what purpose? A primitive hunter may have been adjusted to a life of hunting wild beasts but hardly for work in a twentieth-century city office; similarly, a bank clerk is not well fitted for coalmining, sheep-shearing, mountaineering, or polar exploration. A definition of normality is elusive. For the purposes of this book the average person's expectation of life at birth is thought to be a reasonable yardstick.

Throughout history, man must have become better adapted to his ever-changing environments or else must have adapted his environment immeasurably better to himself. There is either a strong argument for the survival of the physically fitter man or the survival of the civilly fitter society. Either man is constitutionally fitter to survive today or he is mentally fitter, ie better able to organise his civic surroundings. Both conclusions point definitely to an evolutionary progress.

The selection of 'period-pictures' for Chapters 6 to 14 has been a matter of some difficulty and is open to criticism. It is known that there is much in common between one period and another: equally there is a great deal that is different. It is thought that the selection presented offers a reasonable insight into the general evolution of patterns of human disease in Britain from pre-Norman times to the present day.

Notes to this chapter are on p 245.

2

Man in Britain

To examine human health and disease simply in terms of man's relationships with the external environment is like discussing the working of a machine without reference to the material which it will have to process. Environmental influences act on people but their responses, in almost every case, are limited by hereditary or genetic factors. It is necessary to gain some insight into the human material which makes up the population of Britain.

Through the centuries, Britain has been the final landfall and focus of fusion for migrating people from Europe. Not much is known about the earliest settlers of this country except that they came from the Mediterranean. The hunters and fishers of Mesolithic times were few in number and of low stature, averaging 5 foot 3 inches. The first farmers of Neolithic times were also small, with dark hair and long heads similar to many modern Welshmen. After 2000 BC, broad-headed Bronze Age peoples intermixed with them (Fig 2).

Celts began to settle fairly peacefully in the lowlands of southern and eastern England from about 750 to 500 BC after pressures of increasing population had pushed them out of their homelands in the North German Plain and the Rhinelands. A major wave of Celtic immigration to Britain came about 300 BC from northern France and Brittany. These people farmed and traded in southern England and on the hillsides of west and north Britain. A third wave of Celtic peoples came into south-east England after 100 BC. These were Belgic tribes, retreating before the advancing Roman Conquerors of Gaul.

The Celts remained virtually undisturbed in Britain until the main Roman invasion, planned by the Emperor Claudius, in AD 43. In many respects this invasion was only partially successful. Throughout the greater part of what was to become Scotland

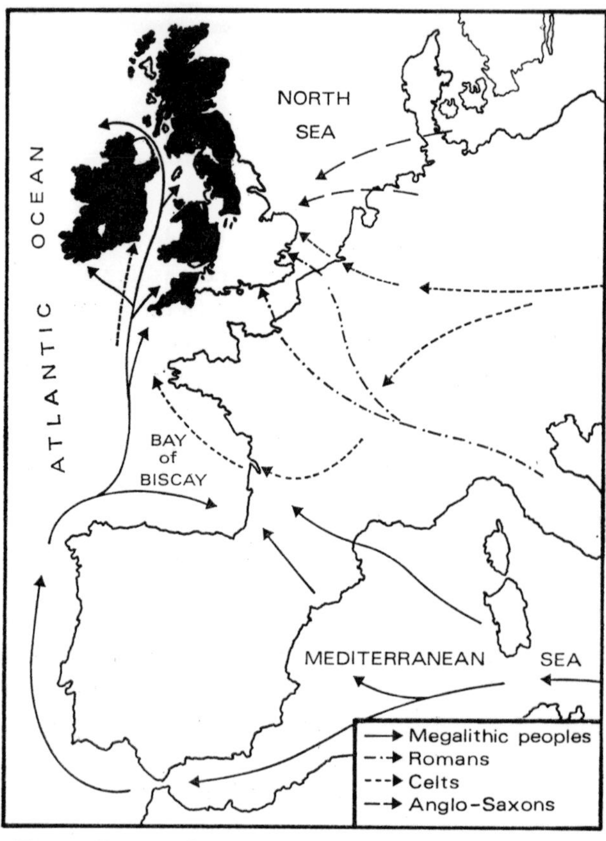

Fig 2 Routes from Europe to Britain followed by early peoples. The shaded areas comprise the 'Highland Zone' of Britain

the indigenous population lived virtually undisturbed during the period of the Roman 'conquest'. This was probably the case in many of the more remote parts of northern England and Wales. Even in the lowlands of south and east England, the Civil Zone of Roman Britain, the population remained basically Celtic. West

Gloucestershire and East Somerset, for instance, were well populated at this time but there were large tracts of the Midlands, particularly Warwickshire, which were very thinly inhabited. The comparatively densely-populated areas of north Kent and the Sussex coast immediately adjoined the Kentish and Sussex Weald, where Romano-British remains hardly occur.

Early in the fifth century the east coast of Britain suffered invasions by Anglo-Saxons and Jutes, tall, mainly fair people, who were basically Germanic. Their descendants are still numerous in south and east Britain, especially in East Anglia and Kent. Movements of colonisers followed these earlier invasions and in time the Anglo-Saxons spread throughout southern England and the Midlands. Their Mercian kingdom expanded in the eighth century to Offa's Dyke.[1] Scotland was left virtually undisturbed by the Saxon invasion.

Spread over northern Ireland, Wales, and much of the north-west of Britain was a warlike people who made fierce raids deep into the Midlands and South. These were the Scots. In the fifth century they spread from Ulster into what is now Argyll and founded the kingdom of Dalriada. In the eighth century the Scots of Dalriada joined with the Caledonians or Picts and founded the territory to which they gave their name.

Norsemen were established in Orkney by the eighth century and from there colonised the north of Scotland and the Hebrides; later they reached the Isle of Man and Ireland and from Ireland spread to south-west Scotland and south to the Cheshire coast. From the ninth century, Danes plundered and later colonised much of eastern England from the Tees to the Thames and set up the Danelaw (Fig 3).

The next and last successful invasion of Britain was made by Normans from France. This meant little more than the replacement of English nobles by Norman nobles and the introduction of French as the language of the upper classes. The feudal system was introduced into south-eastern England and some common grazing lands and forests were enclosed as hunting grounds for nobles.

Environmental contrasts between the 'Highland Zone' in the

north and west of the country and the 'Lowland Zone' of the south and east (Fig 2) played a part in developing important socio-cultural differences in the population of Britain. Many of these differences remain to this day, despite the very considerable

Fig 3 Scandinavian settlement of the eighth, ninth, and tenth centuries—Danes in the east, and Norse by way of the western seas

mobility of the people and the levelling influence of such mass media as the newspaper, radio, and television. The Highland Zone which, in the past, was remote and difficult to invade and conquer, continues to preserve characteristics of the ancient peoples who lived there. It was and still is a region of cultural absorption; the Lowland Zone, easy of access across the Channel, is one of cultural replacement. The people of the Lowland Zone of Britain reflect in large measure the characteristics of succeeding waves of invaders and of the immigrants who have settled there. The south and east of England is still receiving immigrants from Scotland, Ireland, and Wales, displaced people from various parts of Europe, and white and coloured immigrants from such Commonwealth countries as India, Pakistan, Jamaica, Kenya, and Australia. Britain, focus of fusion of Europe's chief racial strains, is as much a creation of European emigration as is the USA.

The ancient patterns of distribution of ancestral populations in Britain have been blurred by movements and intermarriage of people, but the different frequencies of blood groups in the different parts of the country (based on blood-donor evidence) continue to lend support to the above summary of the peopling of these islands.

In respect of certain substances on the red cells, blood may be classified into four types, respectively O, A, B, AB. These blood types are determined by heredity and so their frequencies in a population are a useful pointer to its ancestry. AB is uncommon in Britain, but, with group B, may locally be found in 5 per cent of the people in eastern England and up to 25 per cent in western Britain. Most people have blood of type O or type A. In England south of a line running from north Yorkshire to the Mersey and Severn estuaries and also in south Pembrokeshire, up to 40 per cent of the population has blood of group A (Fig 4). Similar proportions occur in the populations of countries in Europe from which Anglo-Saxon and Viking invaders and settlers came to these parts of England and Wales.[2] Elsewhere in Britain the percentage of the population with blood group A decreases to the west and north.

The O gene is irregularly distributed over much of western Europe but there are some areas with figures of 60–70 per cent. In Scotland there are a few localities where 60 per cent of the population is of blood group O but elsewhere the values are usually between 43 and 50 per cent (Fig 5). Values decrease from west to east and from north to south. High frequencies of O group may indicate survival of characteristics of Neolithic and later prehistoric settlers who came to Britain along the western sea routes. Similar high figures occur among the Icelanders, the Basques, in parts of north-west France and in Sardinia. High

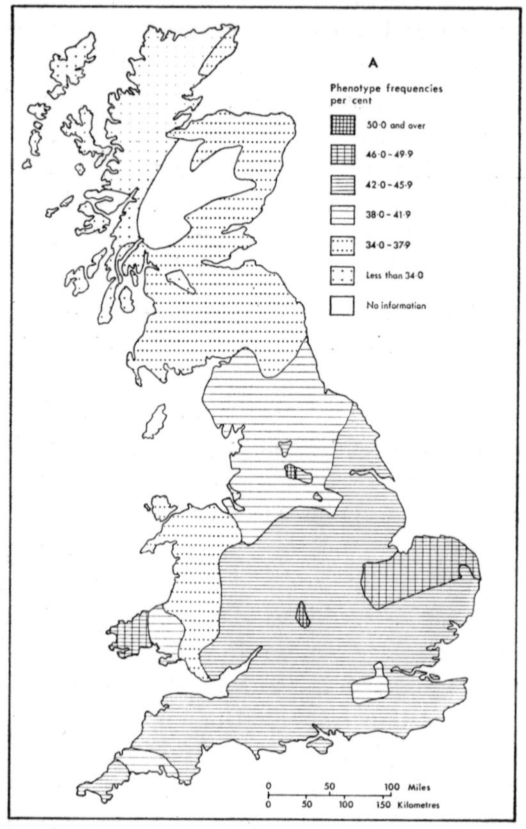

Fig 4 Distribution of blood group A in Britain (*based on Kopec (1971), Brown (1965), and unpublished data supplied by Glasgow and West of Scotland Blood Transfusion Service*)

frequencies of blood group O in southern Scotland and parts of south Lancashire are probably accounted for by relatively recent Irish immigration.

The B blood group gene occurs with frequencies of 5 to 10

per cent over all the British Isles, although there is a rise in frequency in the north and west in more difficult and remote areas which were preferred by prehistoric but not by later peoples

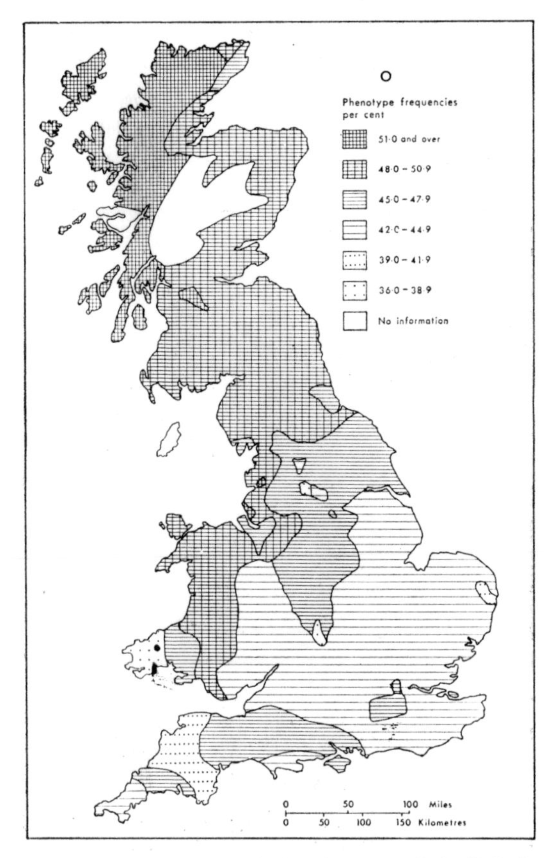

Fig. 5 Distribution of blood group O in Britain (*based on Kopec (1971), Brown (1965), and unpublished data supplied by Glasgow and West of Scotland Blood Transfusion Service*)

(Fig 6). In the interior moorlands of Wales (the Black Mountains of Carmarthenshire, the Plynlimon plateau of Central Wales, and the Hiraethog moorlands of Denbighshire) high frequencies, 16·9 per cent, 13·8 per cent, and 9·5 per cent respectively compared with the national average of 6 to 7 per cent, suggest the survival of characteristics which may go back into prehistory.

In anthropometric terms people in blood group A are usually slightly taller and more round-headed than average, usually with blue eyes and fair pigmentation. Shorter, longer-headed and darker-pigmented people are usually of blood group O.

People belonging to different blood groups appear to differ in their susceptibility to certain of the diseases of adult life. There would seem to be a hereditary predisposition on the part of some

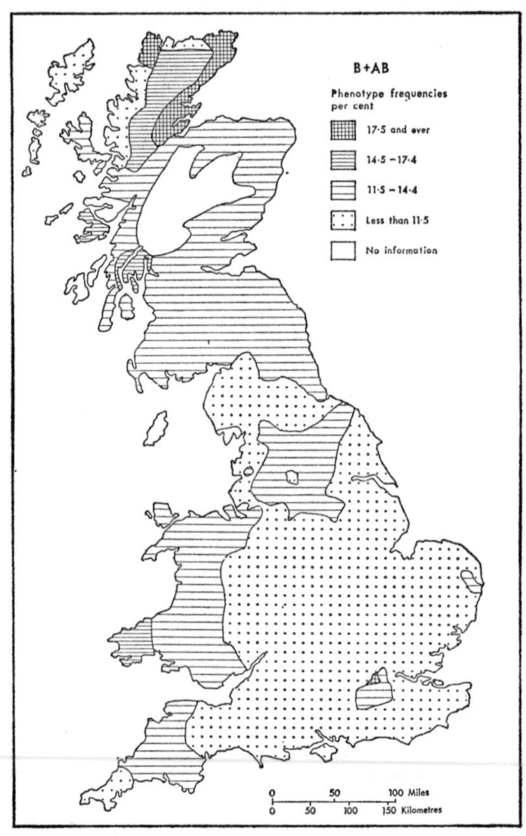

Fig 6 Distribution of blood groups B and AB in Britain (*based on Kopec (1971), Brown (1965), and unpublished data supplied by Glasgow and West of Scotland Blood Transfusion Service*)

people to certain diseases. As noted, disease is rarely the result of one factor; it is influenced by several aspects of both inheritance and environment, by nature and nurture. Illness attributable solely to inherited characteristics, such as haemophilia, is rare, but some associations between the ABO blood groups and disease have been tested beyond all reasonable doubt. The evidence for three associations appears to be overwhelming. The incidence of peptic ulcer is 40 per cent higher in persons of group O than in those belonging to groups A, B, and AB. This increased risk attaching to group O is about 25 per cent commoner in gastric ulcer. It has been established that cancer of the stomach is about 25 per cent more common in persons of group A than in those of other groups. Pernicious anaemia, a disease related to the digestive tract, has a similar group A association. Blood group A also appears to have an association with diabetes.

Table 1 summarises the known claims for disease and blood group associations.

Table 1 Associations between blood group phenotypes and disease

Disease	*Associated ABO or secretor phenotype*
Broncho-pneumonia	A or AB
Bubonic plague	A
Cancer of the cervix	A
Cancer of the pancreas	A
Cancer of the prostate	A
Cancer of the stomach	A
Diabetes mellitus	A
Duodenal ulcer	O
Gastric ulcer	O
Infantile diarrhoea	A
Influenza virus A_2	O
Paralytic poliomyelitis	B (and excess of non-secretors)
Pernicious anaemia	A
Pituitary adenoma	O
Rheumatic fever	AB (and non-secretors)
Salivary gland tumours	A
Smallpox	O
Syphilis	B and AB
Tumours of the ovary	A

Side by side with the positive results have been negative findings for ABO blood groups in the case of a number of other diseases. These include cancers of several sites not listed in Table 1, toxaemia of pregnancy, and hypertension.

It is questionable whether the demonstrated associations are truly causal in the sense that a person of group A is intrinsically more liable to cancer of the stomach or whether persons of other blood groups are protected against the disease. Similarly it may be not that group O carries a liability to peptic ulcer but that other groups have a special protection against it. At present it seems possible that the associations are directly causal,[3] although Weiner is of the opinion that they are almost all fallacious.[4]

Differences in ABO blood groups are at least as old as the human species and the indefinite perpetuation of a polymorphism of this kind must depend on a balance of selective forces favouring the existence of a mixture of different genetic types in the population. The balance or equilibrium is of a dynamic rather than a static kind, an equilibrium moreover which is liable to be changed, albeit slowly, by changes in environmental conditions. Researches into the synthesis of amino acids related to the origins of life have shown that the very existence of the chemicals which influence or rather constitute heredity depend on certain conditions and stimuli. They require the existence of a particular environment. Furthermore, the basic laws of random variation and natural selection ensure that only the creatures that can withstand their environment will pass on their genetic characteristics as only they will survive to do so. Are environment and heredity therefore really irreconcilably opposite poles or are they one and the same thing?

Notes to this chapter are on p 245.

3

Health Hazards of the Physical Environment

The air, water, and land are physical factors of the environment. From the time of Hippocrates, climatic factors have been postulated as influencing, either favourably or unfavourably, man's physical well-being. Well-known British historical treatises in this field include Goad's *Astro-Meteorologica* (1686), Arbuthnot's *An essay concerning the effects of air on human bodies* (1733), Huxham's *Observationes de Aëre et Morbis Epidemica* (1739), and Clark's *The influence of climate in the prevention and cure of chronic diseases* (1830).

The difficulty all along is to isolate from the many components which constitute 'climate' specific factors for detailed analysis. The position of Britain on the planet Earth, ie between latitudes 50°N and 60°N, east of the Atlantic Ocean and west of the great land mass of Eurasia, is the most important factor in both its weather and its climate. The net (*not* prevailing) air movement in these latitudes is from west to east, and since air moving in this direction has normally had a long track over the ocean the climate is generally rainy and equable. There are no marked dry seasons as in, say, tropical or sub-tropical latitudes. Average annual temperatures are relatively high, though they vary appreciably with latitude, altitude, proximity to the sea, exposure, shelter, and other local features (Figs 7 and 8). In the Scilly Isles the average temperature is 8·3°C in January and 16°C in July. London is less equable: the January average is 5°C and the July average is 17·2°C. The more northerly latitude of Aberdeen is reflected in its lower average January (3·3°C) and July (13·9°C) temperatures.

The effects of temperature on the human body are largely a

matter of metabolism and respiratory fatigue. Because internal heat generated by metabolic functions must be dissipated, any impediment to heat loss such as occurs with the high temperatures of tropical climates can depress body functions, lower general vitality, and predispose a person to infectious disease. Conversely,

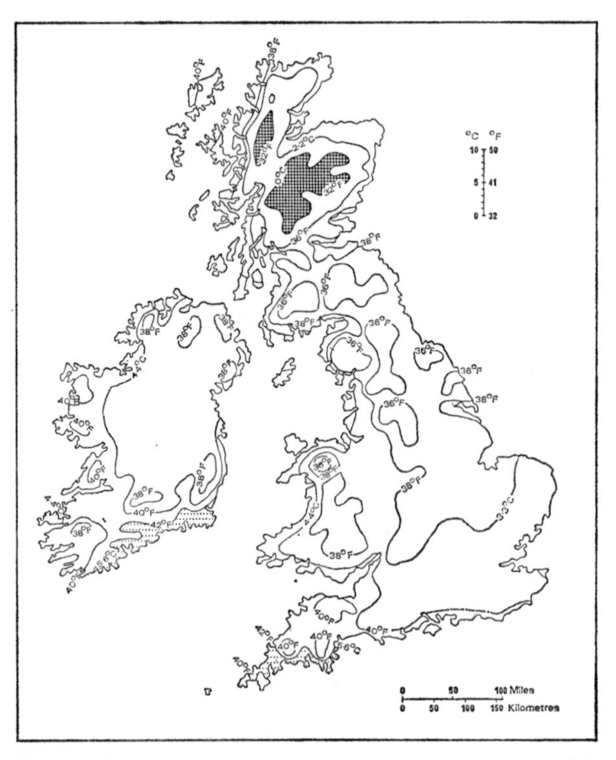

Fig 7 Average actual daily mean temperature for January

greater ease of body heat loss associated with the lower temperatures of temperate latitudes directly stimulates vitality and quickens body functions. Minimum metabolism is observed at 20°C to 25°C. Below 20°C and above 25°C there is a tendency to increased metabolic rate.[1] It is possible to predict with certainty only that extremes of heat and cold are definitely harmful and that moderately hot conditions increase susceptibility to intestinal diseases and moderately cold conditions increase susceptibility to respiratory diseases. Injury produced by excessive heat includes prickly heat, tropical neurasthenia, heat exhaustion, and heat stroke, but these are rare in Britain. Cold injury includes chilblains, frost bite, and hypothermia. British winters are commonly

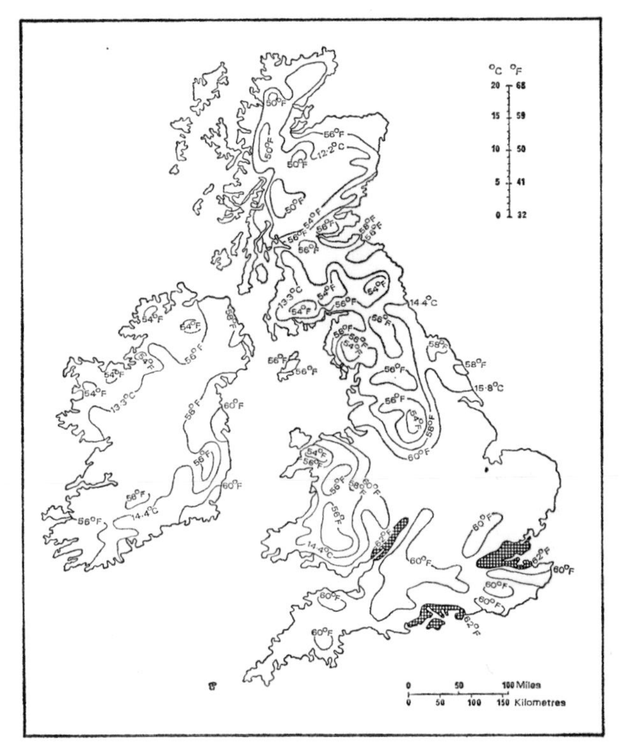

Fig 8 Average actual daily mean temperatures for July

associated with an increase in upper respiratory infections. Cold and lack of natural light curtail many forms of outdoor recreation in winter and there is a tendency for people to crowd together more in places of public entertainment. Natural ventilation in such places, as indeed in the home, factory, office, bus and train, is frequently reduced to conserve heat and at the same time there is a tendency to excessive central or artificial heating. Conditions of crowding, under-ventilation and over-heating are particularly favourable for droplet infections.

The map of mean annual rainfall shows highest falls on the high ground of the north and west of Britain. Similarly, the number of rain days is greater in the west (Fig 9). There are more than 250 rain days a year on the extreme western coast and less than 175 in some parts of the east. Throughout most of the high ground in the north and west the average is over 200 rain days a

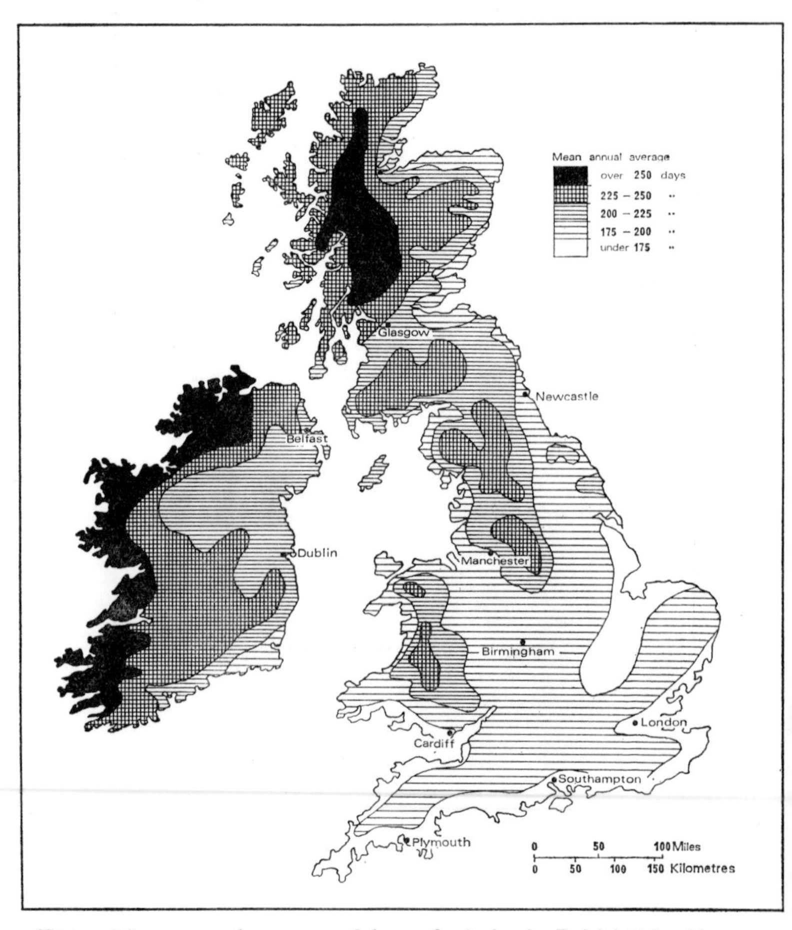

Fig 9 Mean annual average of days of rain in the British Isles (days on which 0·01 inch or 0·25 mm or more of rain is recorded)

year. Popular opinion holds that damp increases individual susceptibility to infection, but volunteers at the Common Cold Research Unit, Salisbury, sitting in draughty passages and wearing damp clothing have been none the worse for the experience. Rheumatic disorders in their several manifestations have long been associated both by the general public and the medical profession with dampness or sudden change of temperature.

It is not generally appreciated that, though the winter half of the year is the wettest in Britain, the *absolute* humidity (ie the amount of water vapour in the atmosphere) is actually at its lowest during this period and at its maximum in July. Waddy[2] has suggested that this seasonal low absolute humidity might exercise a drying action on the mucosa of the upper respiratory tract and lower its vitality and resistance to infection. Boyd[3] found weekly respiratory mortality to be associated equally closely with temperature and absolute humidity, but the very high correlation between these two variables prevents discrimination as to their relative importance.

The qualities which make some parts of the country 'relaxing' and others 'bracing' are not well understood. Wind, humidity, and temperature are probably involved, though the frequency of weather changes locally is another consideration. Fig 10 is a generalised map based on one prepared by Brooks[4] from assessments given by Hawkins[5] and shows those parts of England and Wales which are considered to be tonic and those which are sedative. There are five categories: *very bracing*—mainly limited to the north-east coast with a small outlier in the Peak District of Derbyshire; *bracing*—including the rest of the east coast, part of the Sussex coast, and part of the north coast of Cornwall and hill districts generally; *average*—widespread; *relaxing*—including the south coast of Devon, Cornwall, and Wales and most of the low ground in the western half of England; *very relaxing*—a relatively small area on the western part of the south coast. The main factor in a sedative climate in Britain is the moist equable west or south-west winds from the Atlantic. In Scotland, Galloway, the Glasgow area, the Western Isles, and the immediate west coast have relaxing climates whereas exposure to east winds give Aberdeen,

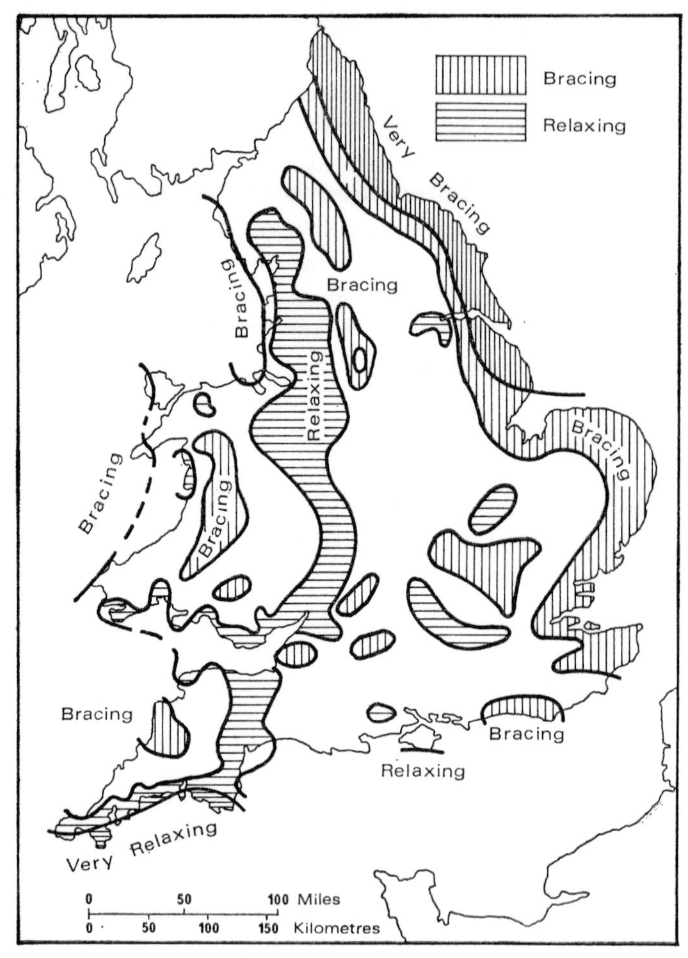

Fig 10 Bracing and relaxing climates in England and Wales (*based on Brooks 1954*)

Dundee, Edinburgh, and the east coast generally a bracing climate. The Southern Uplands, Grampian Mountains, and North-west Highlands have livelier winds and lower temperatures than the neighbouring lowlands and for that reason are bracing.

'Relaxing' and 'bracing' qualities cannot be clearly and definitely differentiated; certainly they are not measurable.[6] The cooling power of the atmosphere, essentially a function of temperature, humidity, and wind speed (their variations rather than their average values) might provide an approximate index of these qualities, with wind probably the most important.

Fig 11 Mean windchill in Britain, January 1956–61. Iso-cooling lines in kg cals-m² hr (*after Howe 1962*)

'Windchill' is a term used to describe the cooling effects of air movement and low temperature, ie the dry convective power of the atmosphere. Since it is a good measure of about 80 per cent of total body heat loss, windchill correlates more closely with sensations of cold than crude temperatures. The chilling produced by a 45 mph (72 kph) wind at −7°C, is about the same as that of a wind moving at 5 mph (8 kph) with a temperature of −30°C.

Fig 11 after Howe[7] shows the distribution of the mean windchill in Britain for January (1956–61) and Fig 12, the distribution of windchill at 18h on 23 December 1961, during an exceptional period of low temperatures and high winds. Exposed human flesh freezes when windchill values exceed 1,400 kg cal m^{-2} hr^{-1}.[8] This figure was attained in Newcastle-upon-Tyne in the early evening of 31 December 1962.

Fig 12 Windchill in the British Isles at 1800 hr, 23 December 1961. Iso-cooling lines in kg cals-m^2 hr (*after Howe 1962*)

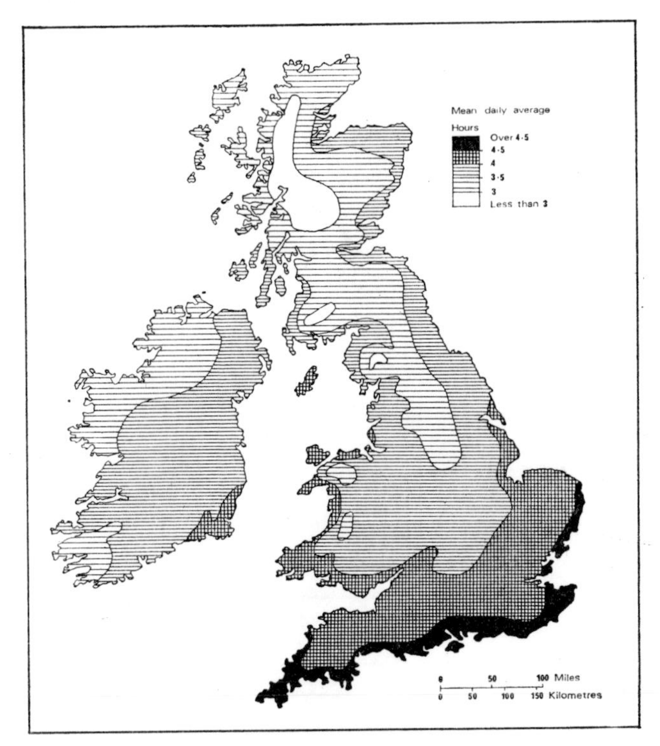

Fig 13 Average number of hours of bright sunshine per day in the British Isles

Wind greatly influences noise dispersion. At the same time it can impair acoustal comfort and provide a degree of nervous irritation even if it is merely audible.

Sunshine causes marked changes in a person's subjective sensations of health. Sick or well people, other than those over-sensitive to sunlight, feel better when the sun is shining. The formation of vitamin D, essential for the prevention of rickets, is stimulated by sunlight (ultra-violet rays) on the skin. Alternatively, excessive exposure to sun during sunbathing can cause skin cancer, particularly those with fair hair and skin (Fig 13).

Fog is the one weather condition in Britain which is undoubtedly and specifically responsible for an increase in morbidity and mortality (Fig 14). A period of fog is always followed by an excessive number of deaths from diseases of the respiratory systems. Three types of natural fog are distinguished:

(*a*) *Radiation fog* which occurs on clear calm evenings and is most frequent in wide damp river valleys such as the Lower Thames, Severn, Trent, or Clyde valleys.

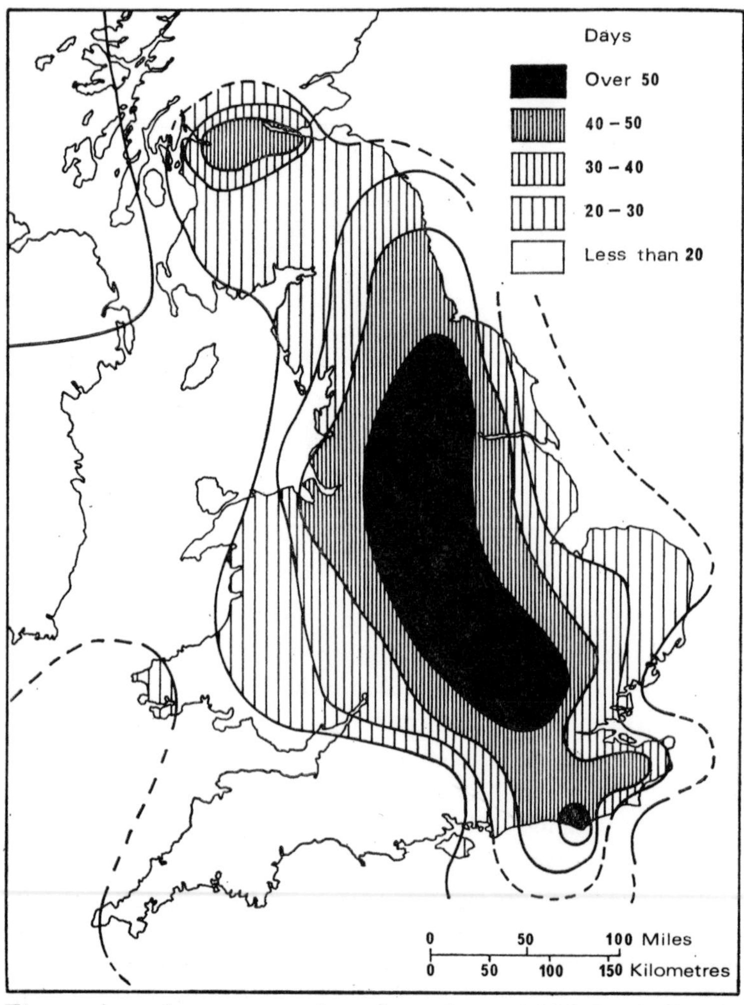

Fig 14 Annual average number of occasions of thick fog in Britain (visibility less than 220 yards) at 0900 hours, 1934–43 (*adapted from Climatological Atlas of British Isles 1952*)

(*b*) *Hill fog* which is simply cloud with base below the level of the highest ground in an area and may form at any time of the day.

(*c*) *Advection fog* which occurs when warm moist air drifts over cold ground. It often occurs along coasts and is generally associated with a thaw after frost.

Until the Industrial Revolution the emphasis was on fog as the embodiment of damp, causing rheumatism, 'agues' and 'fevers', but during the last century or so, town fogs or smog (ie *sm*oke-f*og*) have become synonymous with atmosphere pollution (see p 57).

Climate has undoubted repercussions on the patterns of human disease but the relationship is not straightforward. In Britain which experiences very variable and unpredictable weather conditions, it is the meteorological extreme rather than the mean which is important in disease, especially extremes of solar radiation, temperature, humidity, and wind. The notorious fickle character of Britain's weather arises from the frequency with which the air overlaying the country is replaced. Three main types of air, known as air masses, influence the weather of Britain—maritime tropical masses and maritime polar air masses from the Atlantic which reach the country from the south-west, west, or north-west, and continental polar air masses from Europe (Fig 15). The air masses arrive as air-streams, having acquired their original characteristics in distant source regions. Maritime tropical air masses from beyond the Azores bring warm humid settled weather in summer and unseasonal warmth in winter. When such air masses are cooled by rising over a range of hills they may produce a great deal of rain. Maritime polar air masses from Greenland and North Atlantic are very variable and turbulent, usually moderately warm at ground level but cold at higher altitudes. They are associated with brisk, gusty winds, broken cloud and plenty of sunshine. Rain associated with these air masses comes mostly in heavy showers interspersed with sunny intervals. Continental polar air masses are almost entirely limited to the winter months. They come to Britain from Scandinavia, Eastern Europe or the Soviet Union as biting east or north-east winds. They are infrequent but tend to persist for several days bringing with them dry, bitterly cold conditions. Tropical continental air masses from Southern Europe or North Africa are uncommon in Britain. When present they bring hot dry weather, possibly a heatwave.

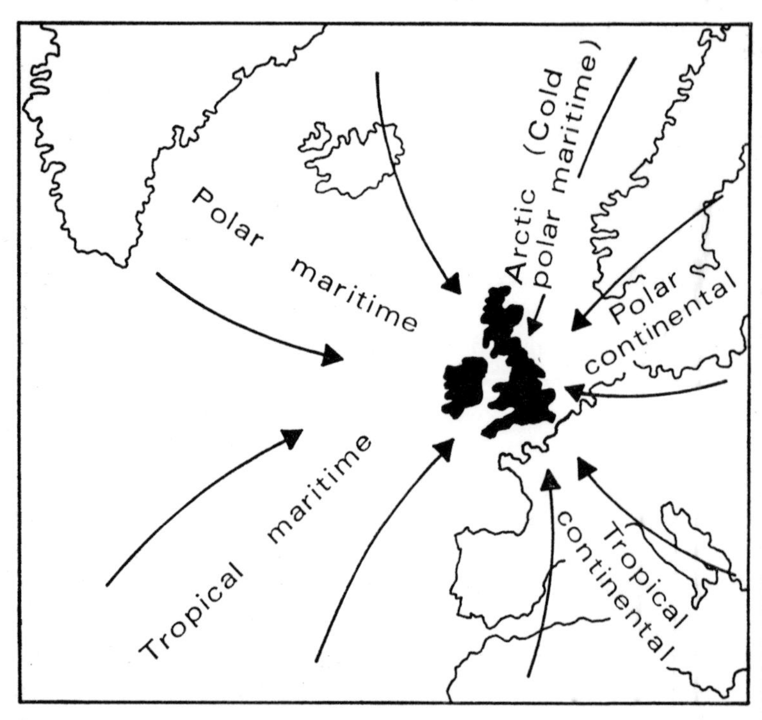

Fig 15 Air masses commonly occurring in Britain

Progressing air masses bring their contrasting weather, though with local variations resulting from relief differences, proximity to coasts, and the presence of urban settlements. Consequently temperature, atmospheric pressure, humidity and wind velocity can vary greatly over short periods of time. Britain is rarely influenced by one air mass at a time. It is usual for the weather to be the result of two or more air masses in conflict, and it is along the boundary or '*front*' between two different air masses in an area of low atmospheric pressure (ie a depression) that weather changes are most marked. Such fronts, especially 'cold fronts, (where a warm air mass is replaced by a colder one), are often accompanied by electrical disturbances. Nervousness, mental unrest, and nervous diseases such as epileptic fits have been reported as increasing on the approach of a cold front. The most extreme form of this effect is the close correlation between the number of suicides and the passage of a front. Some of the observed effects of the approach of disturbed weather, such as sleeplessness, may be due to the effects of rapid fluctuations of barometric pressure acting on the blood. Once again, the extreme form is a proved relation between the frequency of blood clot blockage of the arteries (thrombosis or embolism) and the passage of fronts.[9] Table 2 summarises the reported effects of the influence of weather and climate on diseases.

There are probably few diseases whose distribution is not affected by climate. Apart from the direct action of the elements there is the indirect action since climate and climatic conditions determine in part the nature of the foods eaten, the quality of sanitary methods and appliances, the structure of homes, offices, and factories, social and family organisations, the viability of pathogenic micro-organisms outside the body, the viability and vitality of insects and other animal factors which carry these micro-organisms. These are, in effect, the major direct and indirect agents of disease. It is hardly surprising, therefore, that it is seldom possible to isolate the influence of climate on disease.

Climatic conditions have not been constant throughout the ages. They have changed slightly and subtly since man first arrived in Britain. Such changes may have affected his way of life and the pattern of disease. Prehistoric cultivators arrived in Britain about 3000 BC, when Europe was experiencing a post-glacial climate optimum with temperatures 2°C–3°C higher than now. A gradual deterioration followed which became abrupt about 500 BC, with the setting in of a cool rainy period, though possibly with mild winters. The climate gradually became drier and probably rather warmer during the Roman period until a secondary optimum was reached between AD 800 and 1000. The recording of thirty-eight vineyards in England, in addition to

Table 2 Reported effects of the influence of weather and climate on diseases (abridged from Tromp (1963) and assembled by Maunder (1970))

Short periodical effects	*Long periodical effects (seasonal or pseudo-seasonal)*
Lung diseases	
Tuberculosis: Haemoptysis suddenly increases in clinics after oppressive warm weather before thunderstorms, after föhn, humid cold foggy weather or sudden heatwaves	Increased sensitivity to tuberculin test in March and April; low during autumn
Asthma (bronchial): Increases with sudden cooling (particularly if accompanied by falling barometric pressure and rising wind speed); during high barometric pressure and fog (in W. Europe) very low asthma frequency	Low in winter, suddenly increasing after June, max. in late autumn (W. Europe)
Bronchitis: Increasing complaints during fog (particularly in air-polluted areas) and specially if accompanied by atmospheric cooling	High in winter, low in summer (in W. Europe)
Hay fever (and various forms of rhinitis): Allergic reactions often increase during atmospheric cooling	Hay fever is related to flowering of certain plants or grasses, different for different countries. In W. Europe usually max. complaints in May–June
Cancer	
Skin cancer: More common with increasing number of sun-hours and increased exposure of the skin to the sun	—
Rheumatic diseases	
Most forms of arthritis react on strong cooling (falling temp.; strong wind). Humidity seems to have no direct effect, only indirect through cooling	Arthritic complaints particularly common in autumn and early winter (W. Europe)
Heart diseases	
Coronary thrombosis, Myocardial infarction, and Angina pectoris: Occur more frequently shortly after a period of strong cooling	Highest mortality in Jan–Feb (in W. Europe and northern USA), lowest July–Aug. In hot countries (eg southern USA) highest mortality in summer, lowest in winter
Infectious diseases	
Common cold: Weather changes affecting thermoregulation mechanism, membrane permeability, and growth and transmission of common cold virus seem to initiate the diseases (eg very cold period followed by sudden warming up)	Max. in Feb–March; increasing from Sept–March (in W. Europe)
Influenza: Rel. humidity below 50 per cent and low wind speeds seem to favour the development and transmission of influenza virus	Max. in Dec–Feb; increasing from Sept–March

those of the king, in the Domesday Book (1085)[10] might imply summer temperatures perhaps 1–2°C higher than at present, although there are several vineyards in southern Britain even today. During the period 1250 to 1400 another climatic decline set in, but with a partial recovery from about 1400 to 1550. Between about 1550 and 1850 occurred the so-called Little Ice Age but thereafter, until about the 1930s or 1940s, the climate improved, partly resembling the warmer periods in the Middle Ages. The overall experience of the years since about 1940 tends to suggest that the climate is starting to deteriorate once again. The tendency is to a greater frequency of cooler, wetter summers and colder winters which in turn might influence the future pattern of disease in this country as possibly have past fluctuations of climate.[11] Nevertheless, as noted, it is the occasional climatic extreme rather than the average value which is significant. The frequency with which climatic extremes overlap critical threshold values and trigger off chain reactions, thereby encouraging or inhibiting certain kinds of disease, is of greater importance than any slight climatic change, whether it is amelioration or deterioration.

Apart from its weather and climate, Britain presents an amazing variety of other physical environmental conditions. This is particularly true in the variations in general height and form of the land. Nevertheless a broad dichotomy or dualism is clearly recognisable. The dividing line follows an irregular course from the mouth of the River Tees in the north-east to the north of the River Exe in the south-west. West and north of the line lies Highland Britain; to the east and south is Lowland Britain (Fig 2). In Highland Britain occur the major mountain and hill masses. These cool, humid, wind-swept and water-logged uplands with thin and generally acid soils contrast markedly with conditions to the south and east. In Lowland Britain the land is seldom more than a few hundred feet above sea level and comprises broad plains, low-lying plateaus and scarplands. Conditions on the Mesozoic and Tertiary rocks and the less lime-deficient soils of this part of the country are, in general, more favourable to human settlement than those on the Palaeozoic rocks of Highland Britain.

Differences in background radiation, in the trace-elements in the soil, and in the water supply can be related to differences in basic rock structure, rock type, and general relief of the land. Historical and cultural contrasts between the Highland and Lowland Zone have already been noted.

All forms of radiation, if intense enough, may produce some adverse effects on man, but ionizing radiation is the greatest hazard. Sources of ionizing radiation are principally the members of the uranium and thorium series. They produce a background gamma-radiation and vary markedly from one locality to another. Sedimentary rocks have a lower content of the radioactive elements uranium and thorium and provide less gamma-ray background than igneous rocks. Consequently gamma-ray background from, say, the chalk in Kent or the limestone of the Cotswolds is far less than from the granites of Cornwall or Aberdeenshire (Fig 16). Surveys made in Edinburgh, Dundee, and Aberdeen, where a high proportion of the older houses are constructed from local stone (Lower Carboniferous sandstones, Old Red sandstones, and granites respectively) gave a gonodal dose rate for the population of Aberdeen 20 m.rem or more[12] per year greater than that to the inhabitants of Edinburgh. Recent researches in background radiation suggest that the amounts which people receive are related not so much to local geology but to the type of building material used in houses and other constructions. In some places dwellings may be of local stone but in the majority of cases they are of brick. This is unlikely to be of local origin, or, if it is, is representative of local clay only.[13]

Rocks, together with overlying soils, can have anomolous trace-element or micro-nutrient contents. The essential trace-elements are more important in the nutrition of man than their organic micro-nutrient counterparts, the vitamins. They cannot be synthesised as can the vitamins but must be present in the environment within a relatively narrow range of concentration. Both deficiencies and excesses kill. Soils derive their trace-elements from the soil parent material, applied fertilisers, and agricultural dusts and sprays and pass on their trace-element characteristics to vegetable matter growing in them. Vegetable matter used as food may thus reflect the trace-element peculiarities of the soil and of the parent geological material. An excess of trace-elements such as mercury, lead, cadmium, or selenium whether eaten in vegetable matter or animal foods, can seriously affect health. Deficiencies of elements such as copper, iron, manganese, zinc, iodine, fluorine, cobalt, and molybdenum give rise to nutritional problems[14] (Fig 17).

Trace-element anomalies of rocks and soils may be transferred to the water supply. Rain-water contains small quantities of dissolved atmospheric gases, particularly from oxygen, and weak-

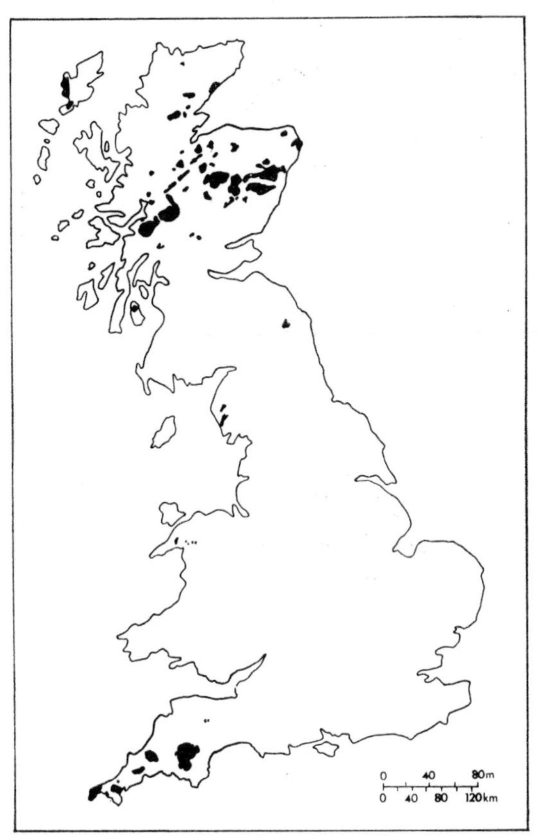

Fig 16 Areas within Britain where the background radiation from solid rocks is relatively high (*S. H. U. Bowie*)

acid from carbon dioxide. Occasionally traces of other acids are present as when nitric acid droplets are formed during thunder storms. In industrial areas, as a result of atmospheric pollutants, rain-water may contain sulphuric acid, organic and inorganic

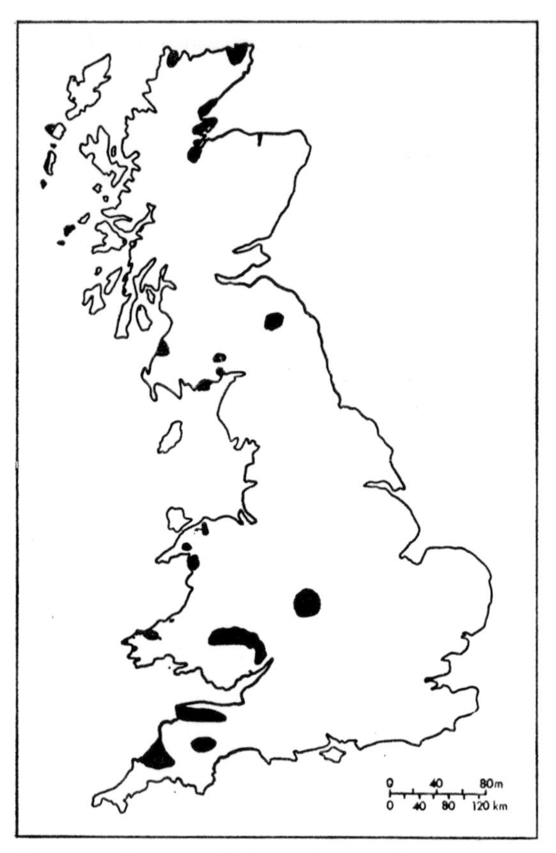

Fig 17 Areas within Britain where cobalt-deficiency soils are known to occur (*Howe 1970*)

dust and other objectionable impurities (Fig 18). Once on the ground rain-water may also pick up humic acids formed by decaying vegetation. Penetrating into the rocks, this acidulated water sets up slow chemical decay of the rock minerals and removes much material in solution.

The quality of the water supply and in particular its degree of

Fig 18 General location of lead–zinc ore fields and defunct mines in the United Kingdom. The inset of part of Central Wales shows the relationship of old mines to streams and rivers which may be chemically polluted by drainage from spoil heaps after heavy rains (*Howe 1960*)

hardness is related to differences in chemical composition of the rocks from which the water is obtained. Hard and very hard water is usually obtained from underground water-bearing formations such as the Chalk, Bunter Sandstone, and Pebble Beds, and the Carboniferous Limestone; soft and moderately soft water comes from surface supplies, particularly from the peat-covered uplands in the west and north of the country. Hardness, due to the presence of sulphates of calcium and magnesium, if present in slight degree, is usually considered good from a hygienic point of view although there is a vague and indefinite association between thyroid abnormalities such as goitre and certain types of hard water (Fig 19). The frequency of occurrence of goitre in the Carboniferous Limestone county of Derbyshire led to its old name of 'Derbyshire neck'.[15]

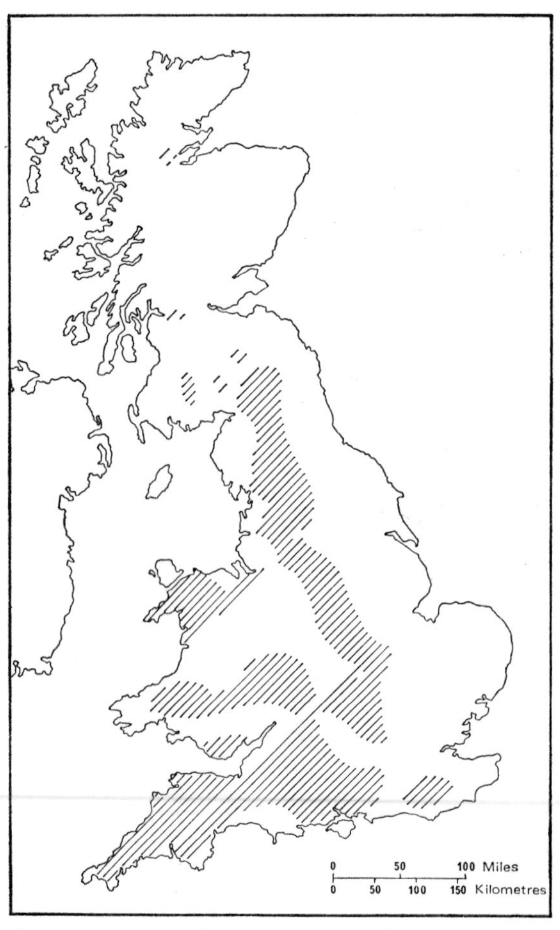

Fig 19 Areas in Britain where endemic goitre has been found (*after WHO 1960*)

Most people appear to prefer soft waters both for drinking and for domestic purposes, yet such waters have been associated with the higher death rates from cardiovascular disease both in Britain[16] and for the USA[17] (*see* p 230). The problem presented here is whether it reflects other factors with which both water, mortality and morbidity are associated. Hard categories of water seem to dominate the area south-east of a line from the Wash to the River Exe. The softer categories are dominant in the south-west peninsula of England, in upland Wales, much of the Pennines and the Lake District and in Scotland (Fig 20). Water may also acquire other chemical characteristics from the rocks through which and over which it flows. Acidity, for instance, will make a solvent for certain metals. The spoil heaps of defunct lead, zinc, and copper mines in Central and North Wales contain residual amounts of these metals, generally as sulphides which are changed to more soluble form by aerial oxidation and the action of acidic waters. Untreated acidic water supplies polluted by effluent from these mines have been implicated in the high incidence of stomach cancer in North and Mid-Wales.[18]

Allen-Price,[19] in a study of variations of cancer deaths in West Devon and in particular the community of Horrabridge, concluded that 'The distribution of cancer in this compact community of Horrabridge which in the radius of a half a mile has three distinct mortality rates, cannot be explained by current theory. Here there is a homogeneous group following the same occupations, eating the same food and in an identical environment, and merely separated one from another by the natural boundaries of the River Walkham which is crossed by a bridge. Here for generations the people have intermarried freely, and their social activities have been combined, yet each artificial section of the community has a widely different cancer mortality. As far as can be assessed, the only difference that could account for this is their water supply.' The high cancer rates were found among that section of the community taking water from wells and springs in highly mineralised strata; cancer rates were low where the water supply came from non-mineralised strata.

Certain mineralised waters, as at Harrogate, Bath, Buxton,

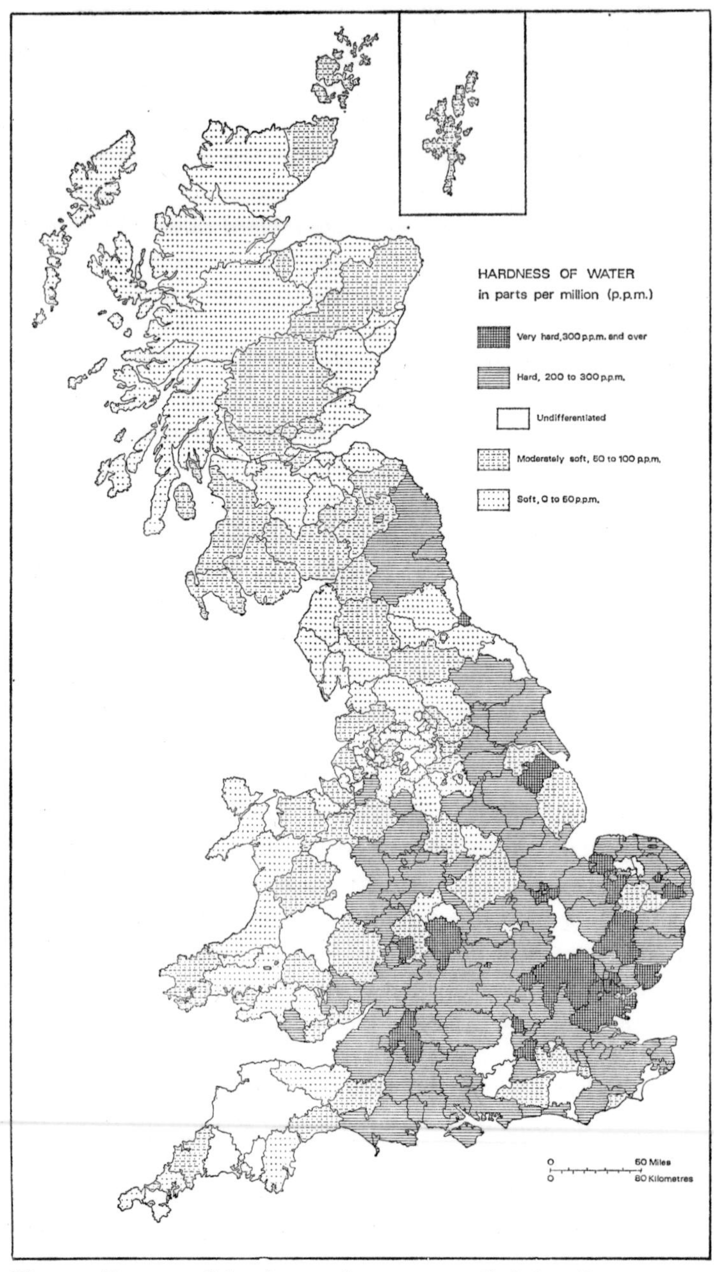

Fig 20 Degree of hardness of water supplied by direct supply water undertakings in Britain in 1968 (*based on data in Water Engineer's Handbook 1968*)

Llandrindod, Strathpeffer, and Bridge of Allan, contain agents which are thought to be of medicinal value. Communities living in areas where the drinking water has a high fluorine content have a low incidence of dental caries.

Health hazards of water are not limited to impurities in solution. Those in suspension—whether living or dead—are also dangerous. Of the living, pathogenic organisms are more important than plankton. The former cause cholera, typhoid, paratyphoid, infectious hepatitis, dysentery, and gastro-enteritis; others are beneficial because they play an essential role in natural purification processes. Dead impurities in suspension may consist of organic matter of decayed leaves, carcasses, animal excrement and certain industrial wastes, or of mineral matter such as fine sand or clay or industrial wastes. Clear sparkling water is often suggestive of contamination by organic matter, opalescent water indicates the presence of colloid material, yellowish water may mean sewage contamination, black-brown water the presence of vegetable matter, and red-brown the presence of iron compounds.

Such is the present demand for water in Britain that the re-use of water is being constantly extended. Two-thirds of London's water comes out of the River Thames at Laleham for purification. At this point the river has already been through the sewage systems of several Thames Valley towns. York drinks water out of the River Ouse after its tributaries have drained a number of North and West Riding towns. Nottingham takes water from the Derwent below the outflow from Derby and a large chemical works. Rivers provide most of Britain's water supply, and in inland communities they take back most of the waste from human bodies, households, and factories. Modern methods of water purification and the capacity of rivers for self-purification make possible the re-use of water, and where water is in short supply, secondhand water is regularly drunk and so far without ill effect (*see* p 56–7).

Notes to this chapter are on pp 246–7.

4

Health Hazards of the Biological Environment

Despite his apparent aloofness from the biological world, man, fundamentally, is still an animal. He occupies a place in the economy of nature and is part of the ecosystem.[1] Man exercises vast influence and much control over his environment but he, in his turn, is influenced by it and, as with other animals, has to contend with natural enemies. He uses many species of plants and animals for his food but in turn his body provides a rich ground for many parasites. Several parasites live in or within man permanently, without causing any structural change or functional disturbance. On the skin there are the staphylococci, in the mouth non-pathogenic strains of streptococci, and in the colon coliform organisms of the bowel, the *E(scherichia) coli*. In the atmosphere are pathogenic bacteria and pathogenic viruses which are responsible for the majority of human diseases. Bacteria are among the smallest living creatures. They are very much smaller than the body cells into which they penetrate or by which they are engulfed. They can be grown outside the body on laboratory media. Viruses are even smaller than bacteria. They are the most minute pathogenic organisms known and the largest can only just be seen under the strongest power of the ordinary microscope. They exhibit the highest known degree of parasitism and cannot live outside living cells.

Most micro-organisms have, in general, an exceptionally active metabolism. Most reproduce by merely dividing into two when they reach a certain size, as do most of the cells of the body. But unlike body cells each of the new daughter cells is capable of further division. Most micro-organisms multiply at a much faster rate than the cells of higher organisms. A typical time for such a doubling is 20 minutes; thus in a single day a bacterium will reproduce 48 to 50 generations. The single bacterium thus could theoretically produce a million million cells a day. If, for instance, there is an infection of a cut by a thousand cells of a virulent streptococcus, these will reproduce something like 10 to 100 million streptococci in 12 hours if left to grow unchecked in the wound and blood. This number is sufficient to kill a man. If the reproduction rate is halved then the number of cells is reduced to nearer several thousand. The significance of bacteriostatic drugs such as the sulphonamides is that they can decrease the reproduction rate of infecting bacteria.

One of the most interesting things about the micro-organisms which attack man is their natural history and the ways in which they are transmitted from one man to another. It is here that the relationship between disease agents and the disease they cause on the one hand, and the physical and human environments on the other, are most clearly seen. Bacteria, spirochaetes, rickettsiae, and viruses may be introduced to the human body directly by inhalation, ingestion, or through abrasions and wounds (as with cholera, epidemic meningitis, tetanus, typhoid fever, and tuberculosis) or indirectly by a carrier or vector (as in the case of malaria, yellow fever, and African sleeping sickness).

For cholera, little more than a name in Britain now but a very serious disease in the nineteenth century, there is a two-factor complex, causative organism and host. The causative organism, the *vibrio cholerae* (discovered by Koch in 1883) is introduced into the human body directly and, as far as is known, only man can be infected by it. Factors or stimuli thought to correlate with, and possibly govern, cholera endemicity are high temperatures, low-lying lands, ponds and lakes, and other bodies of water rich in organic matter and salts, and shelter from the rays of the sun and from rain. These conditions are common in the Indian sub-continent where cholera is thought to have existed since the beginning of recorded history.

Smallpox is another example of the two-factor complex. This

acute infectious disease, caused by the variola virus, arises from either direct or indirect contact with a preceding case of the disease. There are no natural animal carriers or natural propagation of the virus outside the human body; the virus does not live long outside the body, neither is it carried more than a few feet through the air. There were serious epidemics of smallpox in the sixteenth century such as that in 1561–2 (*see* p 114), followed by others at various times during the seventeenth and eighteenth centuries (*see* p 143ff). The disease was present in Britain from 1840–70 but the epidemic of 1900–5 was the last considerable outbreak of smallpox in these islands. The occasional small outbreaks of major smallpox which still occur can usually be traced to infected persons entering this country from India or Pakistan, the main focus area of the disease. If a person with undetected smallpox leaves the Indian subcontinent to travel to Britain by sea visible symptoms are likely to have developed by the end of the voyage. Now, with vastly increased traffic by air, control of the disease is rendered far more difficult since India and Pakistan are little more than ten hours' flying time away from Britain. An infected person can have been incubating the disease in Britain several days before his symptoms become manifest. Modern air travel has made more difficult the prompt detection and isolation of smallpox cases and the subsequent tracing and vaccination of contacts.[2]

Plague, now extinct in Britain, is a disease caused by a bacillus *Pasteurella pestis*. It causes endemic infection in certain rodents and may be transmitted to man by the bite of the rat flea *Xenopsylla cheopis*. It is a complex of three factors: bacillus, rodent, and man. Bubonic plague, the commonest form of the disease, occurs in this way. In human epidemics infection may spread from man to man by coughing up sputum producing a more rapidly fatal form which affects the lung (*see* p 82 *et seq*). The last death of plague in Britain was in 1962 when a worker at the Chemical Defence Experimental Establishment, Porton Down, Wiltshire, became accidentally infected, presumably from laboratory cultures with which he was working.

Under the name 'ague'[3] malaria was endemic in marshy districts of Britain (the Fens in East Anglia, the Isle of Sheppey in the Thames estuary, along the south coast as far as the Isle of Wight and certain other more isolated areas such as the Bridgwater area of Somerset) until the beginning of the twentieth century. The last case of typical English malaria was reported in 1911. Thereafter, for a variety of reasons, indigenous malaria disappeared from Britain. The distribution of malaria and its degree of endemicity are closely related to the distribution of the various species of mosquitoes of the genus *Anopheles* which, in their turn. have different bionomics, ecological characteristics, and breeding habits. The disease is transmitted to humans through the injection of the sporozoites of the malaria parasite of the genus *Plasmodium*. This takes place when an infected mosquito bites a person. Malaria represents a three-factor complex: host (man)—causative organism (*Plasmodium*)—vector (mosquito), the last two having close and direct relationships with water, air temperature, and other climatic conditions. There are plenty of mosquitoes in England capable of transmitting the parasites and there are, at certain times, people in England with malarial parasites in their blood. But the mosquitoes are, so to speak, kept too cold and sporozoites cannot develop in them. Sporozoites do occasionally develop in mosquitoes in England and local people are then infected by malarial parasites taken up by the mosquitoes from the blood of others. This occurred after World War I when many soldiers returned to Britain with malarial parasites in their blood. The species of the parasite that British people are most likely to acquire is *Plasmodium vivax* because the sexual cycle of this species can be completed in the mosquito at lower temperatures than can the sexual cycles of the other species of the parasites. The factors which govern the occurrence of malaria in England have been discussed by McNalty (1943) *et al*, who also examine the factors which govern the geographical distribution of the disease in Holland and other northern European countries.

Typhoid, caused by a bacterium *Salmonella typhosa*, usually enters the body through the mouth in contaminated food, milk, or water. Water contaminated by infected sewage provides a major means of spread. Bovine tuberculosis thrives in cows and

Table 3 Selected transmissible diseases which occur, or have occurred in Britain, together with their biological and other relationships

Disease	*Causative organism*	*Vector*
Cholera	*Vibrio cholerae*	Water carried
Diphtheria	*Corynebacterium diphtheriae*	Direct contact
Dysentery	*Shigella sp.*	Water or food
Hookworm	*Ankylostoma duodenale*	Skin penetration
Malaria	*Plasmodium vivax*	*Anopheles* mosquito
Plague	*Pasteurella pestis*	*Xenopsylla cheopis*
Scarlet fever	*Streptococcus haemolyticus*	Direct contact
Smallpox	*Variola virus*	Direct contact
Syphilis	*Treponema pallidum*	Venereal contact
Tetanus	*Tetanus bacillus*	Penetrating wounds
Typhoid	*Salmonella typhi*	Water and food
Typhus	*Rickettsia prowazekii*	*Pediculus humanus*
Trichinosis	*Trichinella spirelis*	Pork or uncooked sausages
Worm infestation	Tape worms	Beef animals, pork, and fish

may infect a child that drinks milk from an infected cow. Certain loathsome parasitic worms (*Trinchinella spiralis*) live in pigs and can infect people who eat pork sausages and other pig meat which has been insufficiently cooked; others live in dogs and may be conveyed to man if the infected dog licks his hand. The body louse (*Pediculus humanus corporis*) is an important vector of the organisms that cause endemic typhus fever, trench fever, and European relapsing fever. Endemic typhus occurs in people confined in unhygienic and crowded prisons or in armies or among peoples suffering from famine. The lice, infected by feeding upon a person sick with the disease, readily spread under these conditions from one person to another and transmit the causative organism of the disease as they do so.

The distribution of such infectious and parasitic diseases becomes extremely difficult to explain, because the different hosts of the pathogenic viruses, pathogenic bacteria, and parasitic animals may be differently affected by geographical conditions and controls. Furthermore the disease agent itself may be affected directly by these geographical controls when in the alternative host, especially if it is a cold-blooded (poikilothermic) creature such as an insect. And finally, the disease agent, as happens with malaria, may be carried by different vectors in different parts of its range.

Insects are probably the most important disease vectors. They may transmit the disease agent mechanically, as in the case of flies, beetles, and cockroaches, which pollute human food or skin with their feet, saliva, or faeces; or through their bite, as occurs in the case of fleas, lice, ticks, tse-tse flies, and mosquitoes. The

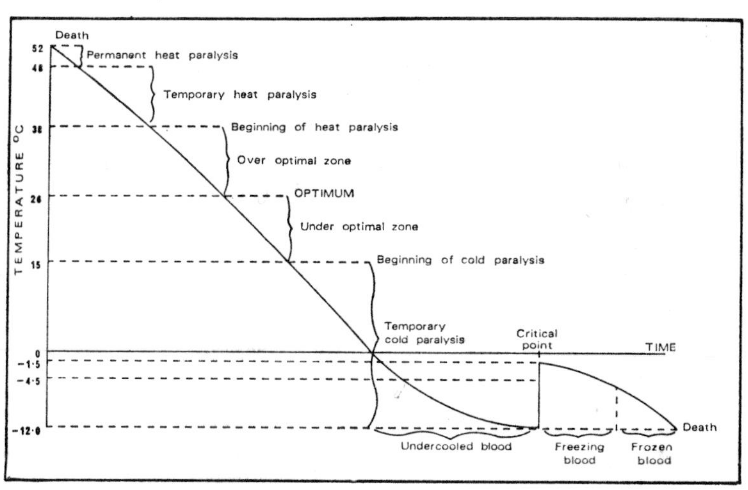

Fig 21 Reaction of insects to temperature (*after Handschin 1928*)

distribution of the different species of vectors is regulated by such critical factors as temperature and humidity. Fig 21 shows the reaction of insects to changes in temperature. In the optimum temperature zone all activities are vigorous. Below this zone lies that of cold stupor, dormancy, and hibernation, and finally death. Above the optimum zone lies that of temporary heat stupor, culminating ultimately in heat paralysis and death as the temperature increases. Not only are insects more active in temperatures between 25°C and 30°C but the processes of their physiological development are accelerated at higher temperatures within these optimum limits, so that certain insects which complete only one life-cycle in a year in cold climates attain several generations in a year in warm climates. Optimum temperatures for insect activity and reproduction are normal in most months in the tropics. Such temperatures are, however, reached only during the summer in Britain. The rapid rate of insect reproduction in the tropics results in several periods of injurious activities by a single species each year whereas in Britain, certain species of insect may reproduce themselves only once a year. Their activities are accordingly restricted, as noted in the case of malaria.

Whether it be the causative organism of a disease (bacterium, spirochaete, rickettsia, virus), intermediate host, or vector, each has its own specific environmental requirements. Each element in a disease complex, including man himself, is inescapably bound up with the geographical environment. Disease in any given locality is the result of a combination of geographical circumstances which bring together disease agent, vector, intermediate host, reservoir, and man at the most auspicious time. Knowledge of these relationships and of each element in the complex is a prerequisite to an understanding of infectious disease, its distribution and control. Furthermore relationships are rarely simple or static. They are highly complicated, far more than the foregoing illustrations suggest or was realised by the early proponents of a 'scientific' germ theory. Pathogenic organisms evolve following mutation or more drastic alterations in genetic constitution. In consequence, descriptions of a disease at the present time may not necessarily conform to the course of that disease throughout history.

The modern attack against microbial and viral diseases centres on either stopping the spread of infecting micro-organisms or interfering with their reproductive potential within a host by means of drugs. The modern antibiotic streptomycin, the synthetic iso-nicotinyl hydrazide, and para amino salicylic acid have proved so successful against tuberculosis that many sanatoria have been closed or used for other purposes. The spread of the causative organism of malaria in the tropics has been restricted by campaigns against the vector *Anopheles* using residual insecticides such as DDT or BHC. Unfortunately certain strains of *Anopheles*

Plate 1 Stoke-on-Trent: (*above*) in 1910; (*below*) in 1969

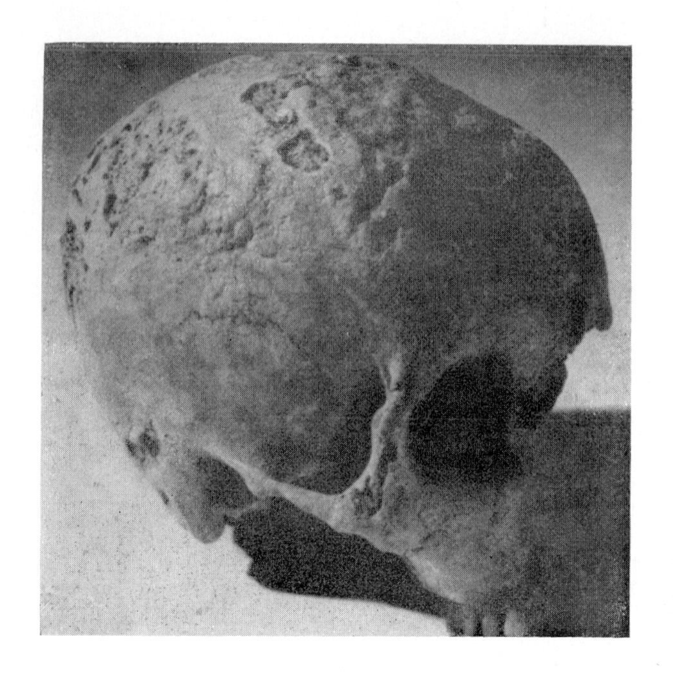

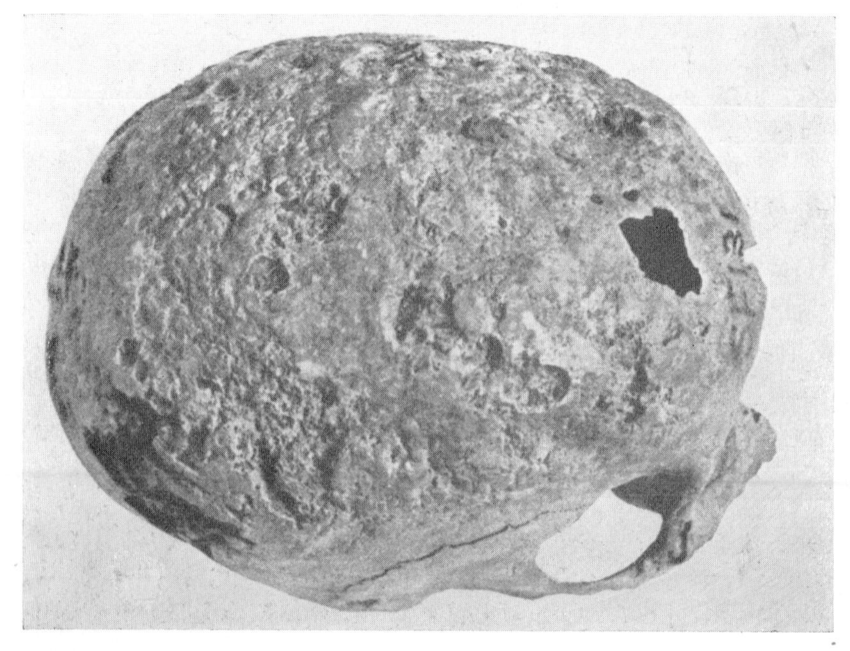

Plate 2 Adult female skull (Saxon) showing evidence (bone erosion and periostitis) suggestive of advanced syphilis

have developed a resistance to DDT and other specific insecticides. One of the most successful alternative insecticides for DDT-resistant mosquitoes has been dieldrin.[4]

Some people are hypersensitive to a wide variety of pollens and in spring and summer suffer from 'hay-fever'. It is said that in Britain each year half a million people suffer from severe 'hay-fever' or similar allergy symptoms caused by grass pollen, dust, or dandruff from pet dogs and cats. Man is also subject to fungus infections. The contagious group includes ringworm and *tinea pedis*, commonly known as athlete's foot.

More recently 'farmer's lung' (*allergic alveolitis*) has made an appearance in this country with symptoms similar to the North American 'maple bark disease' and to the 'paprik-splitter's disease' of Hungary. This is a pulmonary condition resulting from the inhalation of minute particles of mouldy hay or grain, itself a complex material consisting of innumerable fungal spores, hyphae and bacteria, and fragments of vegetable matter. The critical fungal spores are actinomycetes (*Thermopolyspora polyspora*), a simple form of organism which lives in the soil, and whose spores are liberated in dry weather.

In contrast to the health hazards presented by certain aspects of the biological environment, attention might be directed to some of the aspects which are more beneficial. For instance, the large-scale brewing of beers, wine-making, and the making of leaven bread, cheese and certain pickles depend wholly or partly on microbes for their special properties and for the subtle flavours of the resultant foods and beverages. Brewing and bread-making depend on a sophisticated micro-biological technique in which yeast must be preserved from batch to batch. Should this culture be contaminated by certain other bacteria or micro-organisms there will be vinegar instead of beer, or the bread becomes sour and inedible. Microbes are also used to modify natural products occurring in industrial and city wastes. Such wastes are often made less toxic, and some toxic substances may even be converted into useful products.

Notes to this chapter are on pp 247–8.

5

Health Hazards of the Human Environment

Hazards of the human, or socio-cultural, environment, are essentially man-made. They relate to people and population and include the distribution, density, and mobility of population, housing, diet, pollution, agricultural practices, industrial processes and cultural traits.

Prior to the Industrial Revolution, the population of Britain was largely rural and not as mobile as in later years. Industrialisation intensified the process of urbanisation which had been initiated by improvements in agriculture and food storage techniques associated with the 'agricultural revolution'. Urbanisation brought additional disease hazards in its wake. The new industrial towns of nineteenth-century Britain suffered from severe overcrowding and possessed only rudimentary sanitation.

Outbreaks of grave infectious disease such as typhus, cholera, and other bowel complaints, and smallpox, particularly among the child apprentices (some would consider child 'slaves' to be a more appropriate description) in mines and textile mills, were recurrent. So great was the need for labour in the factories that the high death rates put the spotlight on the value of human life. It was Chadwick's *Report*[1] (*see* p 159) in 1842 which precipitated improvements in housing, sanitation, and the provision of clean public water supplies.

Over 80 per cent of the present population of Britain live in towns. Not, however, in the densely-packed, grossly overcrowded dwellings devoid of drainage, water closets, and other basic facilities such as were to be found in nineteenth-century

Liverpool, Glasgow, Manchester, Birmingham, Leeds, Sheffield, Bradford, Nottingham, Newcastle-upon-Tyne, and other big cities (Figs 22 and 23). Such housing has been largely replaced by modern high density local authority houses or multi-storeyed blocks of flats. Air conditioning and central heating of stone, concrete, or brick-built houses and offices provide congenial living and working conditions, in stark contrast to conditions in early Victorian times or to the foul and verminous dwellings of medieval Britain. The artificial interior climates of many of the modern buildings are quite different from the climate out-

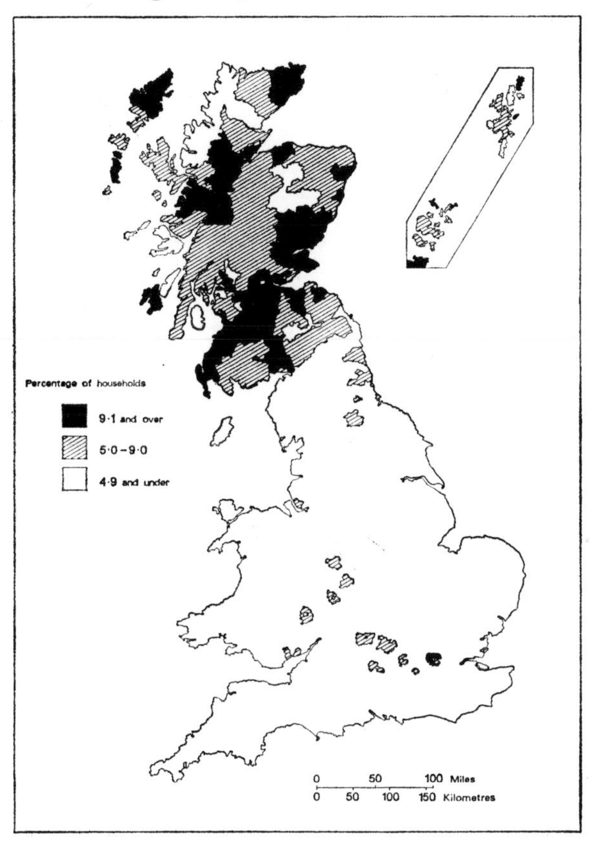

Fig. 22 Overcrowding in Britain, 1961. Percentage of households containing more than one and a half persons to a room. (*Simplified from Lawton 1968*)

side. The large Shell office complex in London, for instance, is designed to accommodate 5,000 people and is completely air conditioned. The heat range is between 20°C and 24°C, individual offices being able to control their temperatures between these figures in the winter while in summer the system is under cooling load. Relative humidity is maintained between 40 and 60 per cent. It is difficult to match this 'climate' among the natural climates of the world because it is equable and the relative humidity is low. It might prove detrimental to health because of its monotony if the people concerned lived in it for the whole of

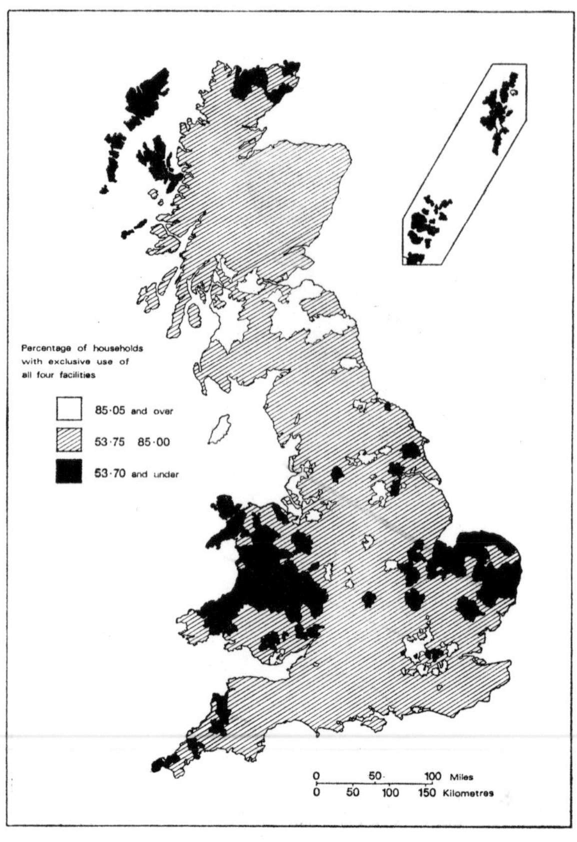

Fig. 23 Household facilities in Britain, 1961. The facilities comprise cold and hot water taps, a fixed bath and a water closet. (*Simplified from Humphrys 1968*)

their time rather than the working day. A fairly close approximation is offered by leeward regions in the Hawaiian Islands. Whether such artificial climates are optimal for a person's physical or mental functioning is not known. It is thought, however, that air conditioning provides relative freedom from infection for the occupants of buildings where it is installed, due both to the withdrawal of infected air and the filtration of incoming air.

Food is an important factor which can provide hazards to health. Over the centuries there has been a change in food and food habits, and in particular, a reduction in the amount of protein and an increase in carbohydrates. Some authorities would argue that compared with the total evolutionary history the relatively short time since man changed from a protein-rich diet to a carbohydrate-rich diet has not permitted adaptation.[2] This being the case it seems unlikely that the people of Britain are as yet fully adapted to some of their contemporary diets. Sugar is used in ever-increasing quantities (Table 4) and chemicals are added to food and drink to improve palatability and appearance. Cyclamates, derived from benzine, long used as a substitute for sugar, have now been banned, as have many artificial colourings.[3]

During World War II, when food rationing was enforced, the findings of nutritional science were applied to the task of feeding the population. It became necessary to provide bread of high nutritive value, to increase the consumption of potatoes, oatmeal, cheese and green vegetables, to supply not less than a pint of milk per day to expectant and nursing mothers, and to all children up to the age of fifteen years, and to fortify margarine with vitamins A and D, in order to maintain a balanced diet. This was achieved. At the end of the war the national diet suffered temporarily following the termination of American Lend-Lease but since about 1953 it has been of a very high order. There are, however, interesting regional variations (Fig 24) with household consumption in the London area well in excess of the national average for all selected items of food except margarine, cakes, and biscuits. Consumption in Scotland is well *below* the average for the same items except margarine, cakes and biscuits, and beef. Wales is characterised by high consumption of butter, the

Table 4 Trends in United Kingdom food supplies from 1880 to 1965 in pounds per head per year (*from Greaves and Hollingsworth 1966*)

	1880	*1909–13*	*1924–8*	*1934–8*	*1941*	*1944*	*1947*	*1950*	*1953*	*1956*	*1959*	*1962*	*1965*
Liquid milk	213	219	217	217	265	308	303	345	330	323	319	325	325
Sugar	64	79	87	96	67	71	82	84	98	109	111	111	110
Butter	12	16	16	25	10	8	11	17	13	16	19	20	20
Margarine	0	6	12	9	18	18	15	17	18	17	14	11	10
Fish	18	41	41	26	16	20	32	22	20	22	22	21	21
Meat (carcass weight including bacon and ham)	91	131	129	129	99	110	96	112	111	134	132	142	135
Potatoes	296	243	230	190	188	275	286	242	245	225	211	214	213
Wheat flour	280	211	198	195	237	234	225	206	193	179	168	161	155

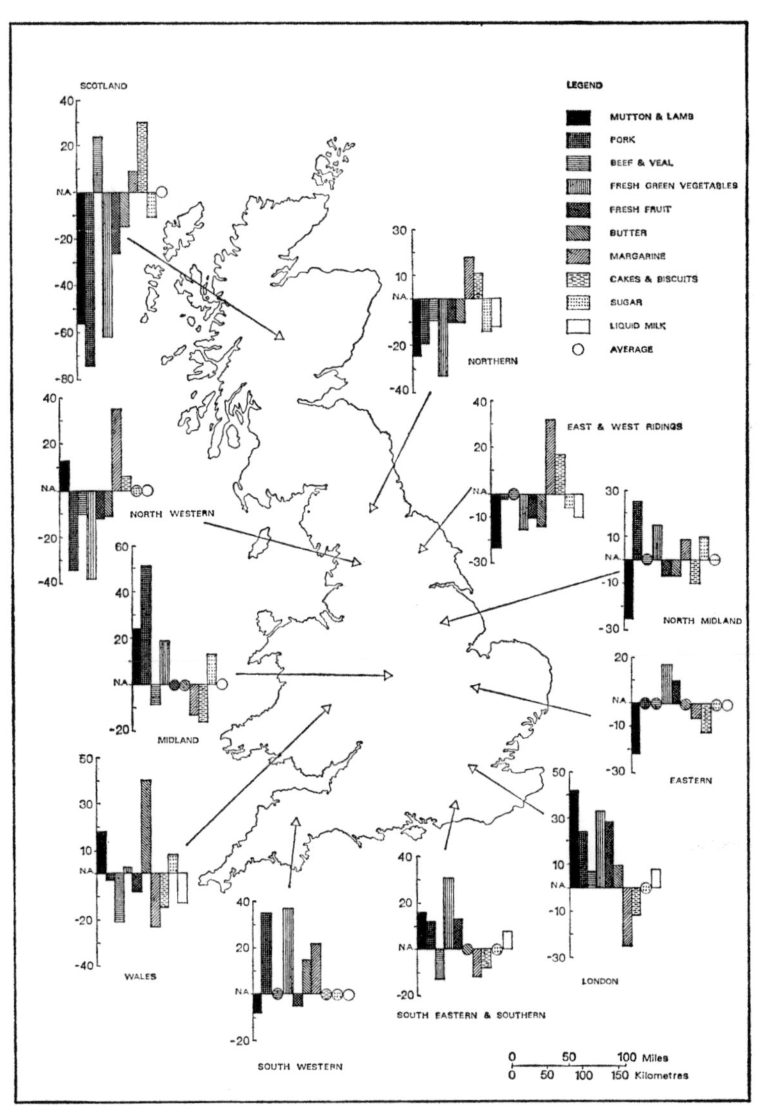

Fig 24 Regional variation in household consumption of selected foods in Britain, 1960–3. The horizontal lines marked NA represent the national average consumption of the selected foods. (*Based on Annual Reports, National Food Survey Committee*)

north-west and north-east of England for high consumption of margarine, and the Midlands for pork.

Battery and deep litter systems of egg and poultry production, the 'sweat box' system of rearing pigs, 'barley-beef' and the intensive systems of livestock husbandry and other methods of food production contrast markedly with the more traditional methods. The use of drugs, growth stimulants, and other food additives to the food of livestock reared under intensive methods can introduce unsuspected residual hazards and possible adverse consequences for health. The lacing of young animal's food with traces of broad-spectrum antibiotics has been such that Britain's cattle and fowl stocks have become reservoirs of drug-resistant germs.[4] Chloramphenicol, the only drug effective against typhoid, could become useless because of the vogue for giving it to broiler fowls.

There is now a wide variety of foods at the disposal of the population, to be selected as required or desired. Canning, chilling, freezing, dehydration, and other developments in food technology help to satisfy practically every need, and rarely is the nutritional value impaired.

Some changes in farm practices must not pass unnoticed in the context of man-made health hazards. Soil, a complex system of mineral, chemical, and biological components, is the substance in which plants grow and is the main source of food for man and his animals. The maintenance of soil fertility is a basic requirement for the continuation of human life and is achieved through the combination of the application of farmyard manure, compost, mineral matter, or the practice of crop rotations. Chemical fertilisers such as superphosphates and ammonium nitrates are being applied to the soil in increasing amounts. The application of such chemical fertilisers causes entroplication of near-by waters and affects soil structure, but has no deleterious effect. On the other hand, the unrestricted use of powerful chemical pesticides based on organo-phosphorus compounds and chlorinated hydrocarbons, while revolutionising chemical warfare against harmful insects and pests, upsets the balance of the soil ecosystem and may prove a serious source of contamination of food supplies (Fig 25).

Annual Reports of the Nature Conservancy have disclosed that poisonous residues from agricultural pesticides occur in the eggs of seabirds collected over widely separated places around the coasts of Britain. If the coastal waters around Britain are substantially contaminated by pesticides it is clear that the soil is contaminated, and probably to a far greater extent than is generally appreciated.

The modern town, the social habitat of industrialised man, is characterised by high-density living and overcrowding, by an atmosphere polluted by diesel fumes, smoke and sulphur dioxide,

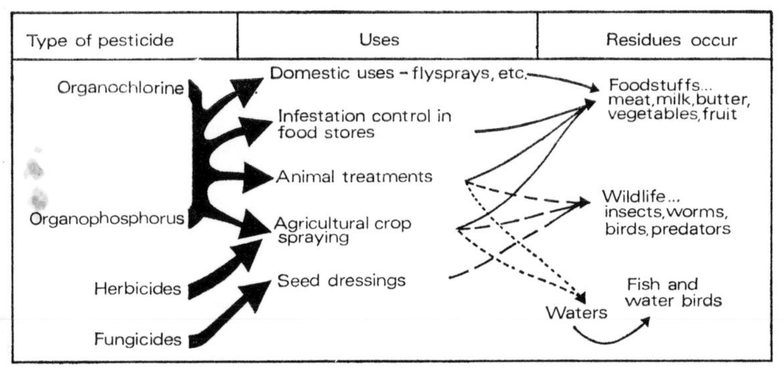

Fig 25 Pesticides, their uses and the location of residues (*after Abbott and Thomson 1968*)

and by the noise of heavy traffic. The pollution over cities and industrial areas, grit, dust, smoke (80 per cent from domestic chimneys) and gases such as sulphur dioxide, carbon monoxide, hydrocarbons, and fluorine (Table 5) was, until the introduction of smoke-free zones in cities in 1956, sufficient to reduce the duration of sunshine in such cities as Leeds, Sheffield, and Manchester in winter months to less than half that in outlying districts. The reduction in average bright sunshine in central London compared with areas outside the metropolis amounted to a loss of 44 minutes a day. In Glasgow the loss in the centre was 20 minutes compared with the western outskirts. The introduction of smoke-free zones in the cities of Britain has been most successful and given rise to a far clearer atmosphere. London and other large

Table 5 Atmospheric pollutants and their sources

Atmospheric pollutants	*Main source*	*Diseases or disorders commonly associated with them or thought to be caused by them*
Beryllium	Beryllium extraction plants	Granuloma of the lung (beryllium lung disease)
Carbon monoxide	Chiefly exhausts from motor vehicles (not diesel-engined vehicles)	Aggravation of respiratory disorders
Fluorides	Enamelling works, some steel plants, artificial fertiliser plants, bauxite processing, pottery kilns, brickworks, aluminium smelters	Fluoridosis in animals grazing on contaminated pastures: irritation of skin and mucous membranes in man
Hydrocarbons	Oil refineries and exhaust gases (unburned petrol) from motor vehicles	Cancer of the lung
Inhalable dust, SO_2	Coal and oil heating	Bronchitis and/or irritation of mucous membranes of respiratory system
Ozone	Not ascertained—probably interaction of ultra-violet rays and products of combustion in presence of a catalyst	Bronchitis and/or irritation of mucous membranes of respiratory system
Pollen	Ragweed	Hay fever
Polcyclic organic compounds (especially 3,4-benz-pyrene), also certain aliphatic hydrocarbons	Incomplete combustion of hydrocarbons	Factor in cancer of the lung

cities are now far less smoky, sooty, and foggy than they were fifteen or more years ago (plate p 47).

The average concentration of smoke and sulphur dioxide in urban areas throughout Britain is about 200 microgrammes per cubic metre, although this may vary by some 15 per cent according to changes in weather conditions from one year to the next (Figs 26 and 27). Concentrations vary little throughout the year, but on occasional days in winter levels may be three, ten, or twenty times greater than average. These are usually 'smog' days when, as a result of 'temperature inversion'[5] the upward drift of chimney products is checked and the intensity of the impurities increases in stagnant air near the ground.

The classic example was 'the great smog' which engulfed London from 6 to 10 December 1952. On that occasion, sulphur dioxide levels at Monk Street in the City of Westminster and in Golden Lane in the City of London reached 7,000 or 8,000 microgrammes per cubic metre. Although at first the dire results of this massive pollution were not appreciated, it is now thought to have caused the deaths of 3,500 to 4,000 persons from chronic bronchitis.

There are pollutants of more local importance, associated with industrial processes in manufacturing industry and of domestic

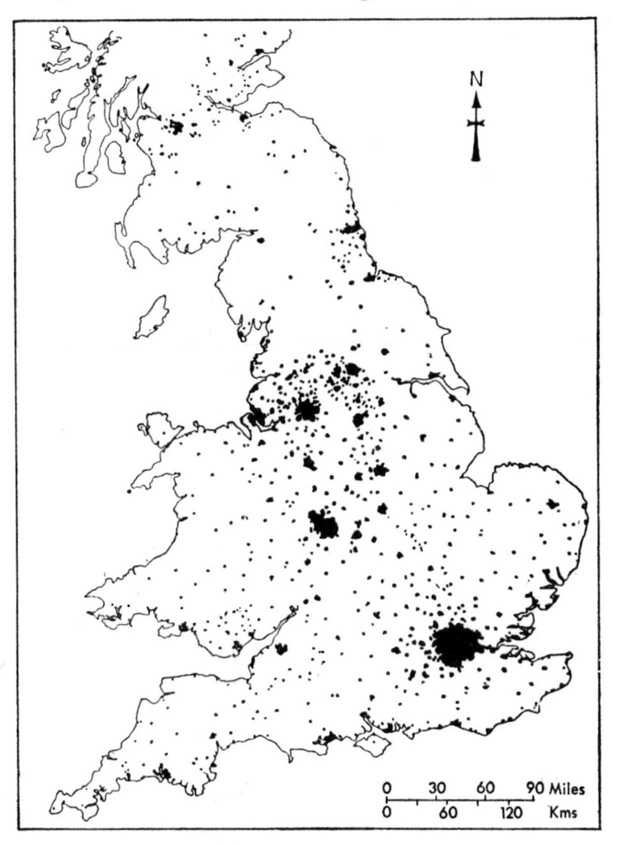

Fig. 26 Distribution of smoke in Britain. Areas shown in black have an average concentration more than 100 micro grammes per cubic metre (after *Dept Sci Ind Res*, 1960)

importance associated with vaporising fly-killers, washing-up detergents, and weedkillers. These present a range of health hazards in the form of absorption or contact with poisonous or deleterious substances. As early as 1775 Percivall Pott drew attention to soot as a cause of scrotal cancer in chimney sweeps. Silicosis is a risk in quarrying and glass manufacture and there is an above-average incidence of pneumoconiosis among coal-miners. Lead, mercury, arsenic, fluoride, chromium, and benzene are among the recognised poisonous materials used in modern industry. These, and hundreds more new chemicals, are being

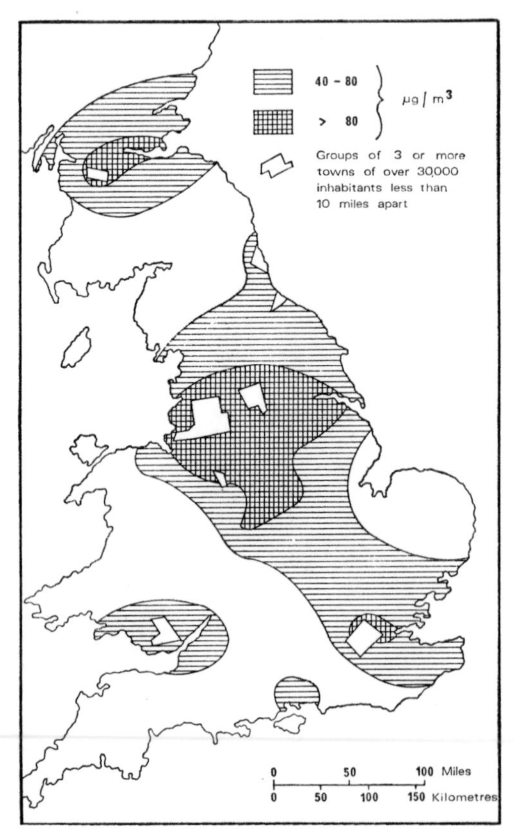

Fig 27 Distribution of sulphur dioxide in Britain. Winter mean values in surface country air, mg/(1000 m³). (*After Meetham 1964*)

introduced into the environment each year. Diseases or disorders commonly associated with atmospheric pollutants and thought to be aggravated by them include chronic bronchitis, pneumonia, lung cancer, emphysemia, and asbestosis.

In contrast to former times, the tempo and tensions of life in the large modern urban communities of Britain are such as to lead to 'stress' believed by some physicians to be a factor in producing coronary heart disease, cerebro-vascular disease ('stroke') and some cancers. Deaths from these causes are considered to be premature and give cause for concern. How much mental illness is due to the stresses of modern life, to genetic causes, or to influences in early years in the Freudian sense is still unknown. Noise in towns from motor vehicles, jet aircraft, building site machinery, pneumatic drills, and, to a lesser extent, ice-cream and other vendors' chimes, television sets, record players, and tape recorders, is a contributory cause of serious nervous disease (Fig 28). The noise of winds in 'skyscraper flats' produces tension from fear of actual physical harm. On the other hand, the quiet of the more spacious new towns and low-density suburbs sometimes causes stress among those not accustomed to it.

Increasing use of water closets in the nineteenth century instead of the earth midden led to the direct pollution of streams, rivers, and coasts with sewage. Indeed the condition of the water and foreshore of the Thames in the middle of London in the nineteenth century was so foul that sheets soaked in disinfectant (chloride of lime) were hung in the Houses of Parliament in an attempt to counteract the stench (plate p 203).[6] Water was, and is, also contaminated from other sources. In particular, wells are subject to seepage from manure-heaps, cess-pits or the more recent septic tank.

Sanitation in most parts of Britain is now of a far higher standard than in the last century and, except for towns and cities near the sea, sewage is generally treated and neutralised before it is disposed of. But rivers, and ultimately the sea, continue to function as the cheapest industrial lavatory in the world. In addition to sewage, industrial effluents including such toxic substances as phenols, oils, and detergents are discharged into streams and rivers. Pulping plants release sulphate and kier liquors and mines pump out highly mineralised waters containing sulphates, chlorides, and calcium and magnesium compounds. A survey of rivers in England and Wales, conducted by the

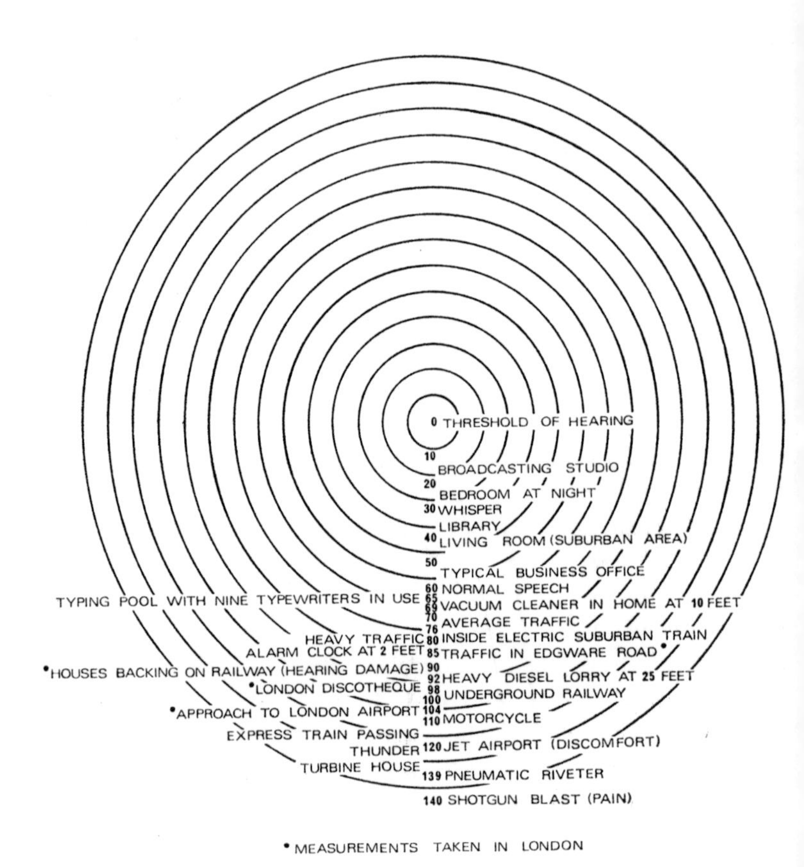

Fig 28 Sound levels in decibels

Ministry of Housing and Local Government, estimated that in rivers with a dry weather flow of over one million gallons a day, about 12 per cent of the total mileage treated was so badly polluted as to be in urgent need of further treatment and a further 15 per cent was of doubtful quality. Unfortunately the most intensely de-oxygenised waters occur in the middle and lower sections of rivers running through populated areas. In 1969 the River Tame had a fair claim to be the filthiest river in Britain. Fifteen million gallons of virtually untreated sewage from Birmingham and three other West Midlands boroughs debouched daily into the Tame, a tributary of the Trent. Not even the log-louse and the blood-worm, the lowest forms of river creature, could sustain life in the slate-grey, opaque, and putrid waters of the river. The usual water purification processes are virtually useless for many of the pollutants in industrial effluent. There is now the additional problem of disposing of radioactive waste from nuclear power stations and plutonium plants (Fig 29). The long-term effect and chain reaction on man of the accelerated life-cycle of some marine organisms in the high temperature sea water near such nuclear stations and plutonium plants is as yet unknown.

Nuclear explosions and radiations are potentially the most decisive of all events in human history in their significance for environmental change and predisposition to disease in man. Radiations from rocks and cosmic sources constitute a perfectly natural part of man's environment and vary from one part of the country to another. Human population have adapted themselves to such background radioactivity and to its regional variations. Now, however, radioactivity is likely to increase as a result of the explosion of nuclear bombs. Nuclear fission and fusion release large quantities of radioactive isotopes into the atmosphere. Strontium-90 is probably the most dangerous of the materials resulting from nuclear fission. It is a bone-seeking isotope which extends down from the atmosphere to the soil and to the sea. It is then taken up by plants, growing organisms, plants, birds, animals, marine plankton, and fish, and thereby enters into ecological situations everywhere. Caesium-137 is another dangerous radionuclide. Through external exposure and food such isotopes present a new and dangerous hazard to man's health and genetic stability. Some reassurance might be gained from a recent report which states that the amount of Strontium-90 deposited on the earth has been falling steadily since 1963 when the last large-scale tests finished. However the report was prepared before the

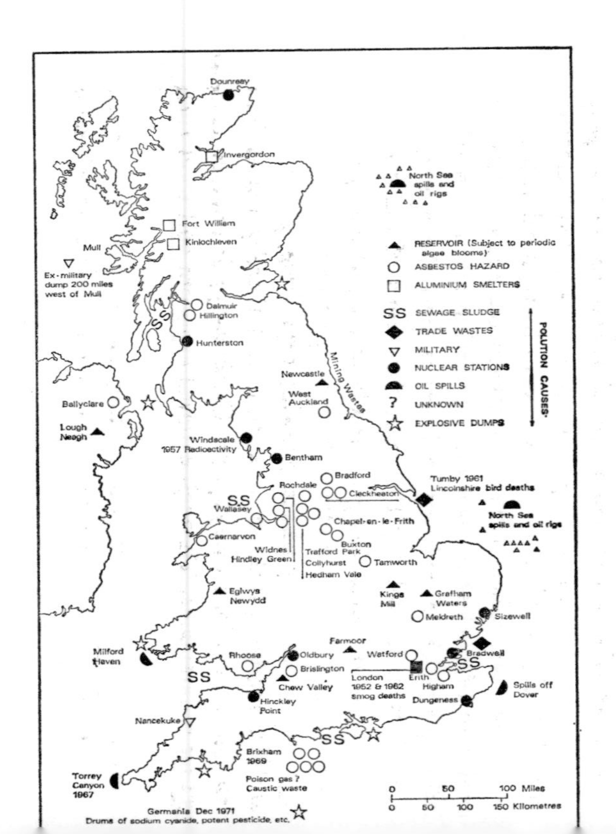
Dounreay
Invergordon
North Sea spills and oil rigs
Fort William
Kinlochleven
Mull
Ex-military dump 200 miles west of Mull
RESERVOIR (Subject to periodic algae blooms)
ASBESTOS HAZARD
ALUMINIUM SMELTERS
SS SEWAGE SLUDGE
TRADE WASTES
MILITARY
NUCLEAR STATIONS
OIL SPILLS
? UNKNOWN
EXPLOSIVE DUMPS
POLUTION CAUSES
Dalmuir
Hillington
Hunterston
Newcastle
West Auckland
Mining Wastes
Ballyclare
Lough Neagh
Windscale 1957 Radioactivity
Bentham
Bradford
Rochdale
Cleckheaton
Tumby 1961 Lincolnshire bird deaths
North Sea spills and oil rigs
SS
Wallasey
Chapel-en-le-Frith
Caernarvon
Buxton
Widnes
Hindley Green
Trafford Park
Collyhurst
Hedham Vale
Tamworth
Eglwys Newydd
Kings Mill
Grafham Waters
Meldreth
Sizewell
Farmoor
Milford Haven
Rhoose
Oldbury
Watford
Bradwell
Brislington
SS
Chew Valley
London 1952 & 1962 smog deaths
Erith
Higham
Spills off Dover
Hinckley Point
Dungeness
Nancekuke
Brixham 1969
Poison gas?
Caustic waste
Torrey Canyon 1967
Germania Dec 1971
Drums of sodium cyanide, potent pesticide, etc.
0 50 100 Miles
0 50 100 150 Kilometres

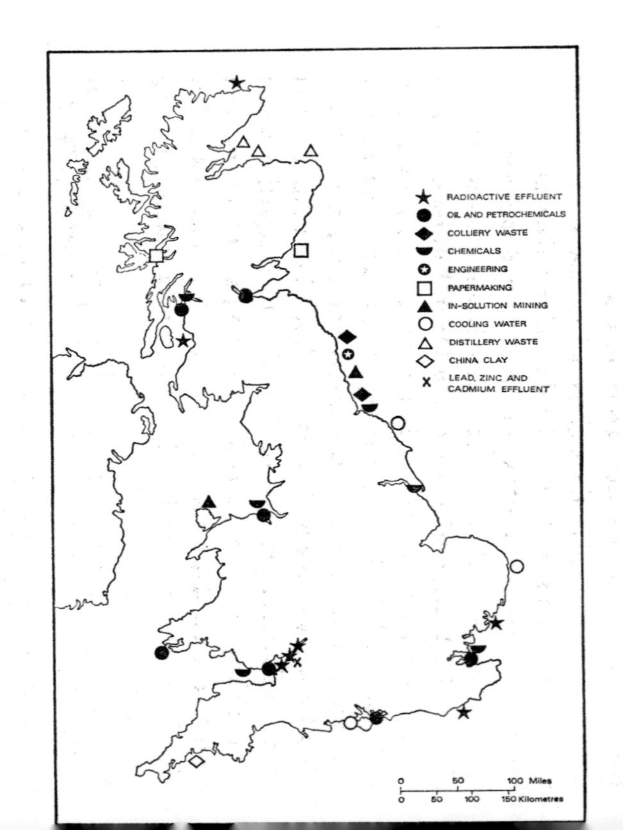
RADIOACTIVE EFFLUENT
OIL AND PETROCHEMICALS
COLLIERY WASTE
CHEMICALS
ENGINEERING
PAPERMAKING
IN-SOLUTION MINING
COOLING WATER
DISTILLERY WASTE
CHINA CLAY
LEAD, ZINC AND CADMIUM EFFLUENT
0 50 100 Miles
0 50 100 150 Kilometres

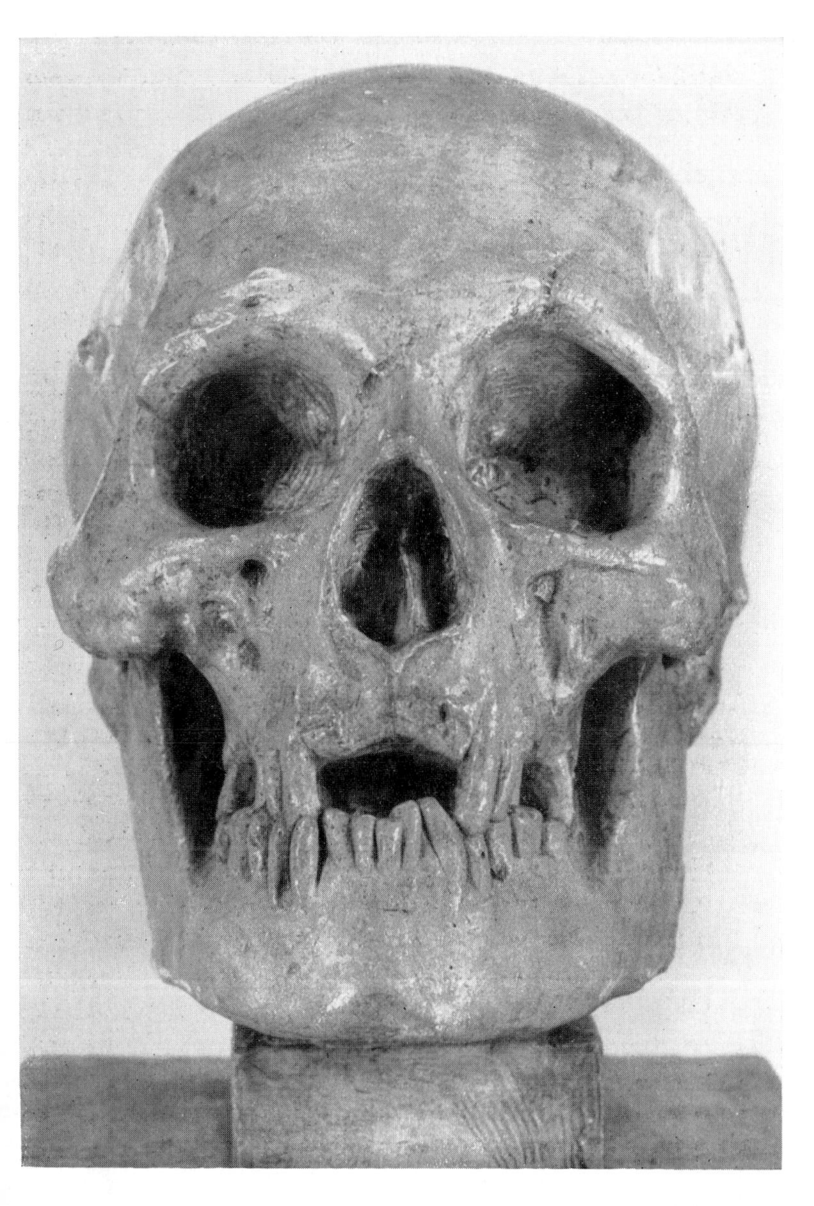

Plate 3 Skull of King Robert I 'The Bruce' (1274–1329) of Scotland, showing evidence (bone erosion and destruction) suggestive of leprosy

The Diſeaſes and Caſualties this Week.

Abortive	4
Aged	45
Bleeding	1
Broken legge	1
Broke her ſkull by a fall in the ſtreet at St. Mary VVool-church	1
Childbed	28
Chriſomes	9
Conſumption	126
Convulſion	89
Cough	1
Dropſie	53
Feaver	348
Flox and Small-pox	11
Flux	1
Frighted	2
Gowt	1
Grief	3
Griping in the Guts	79
Head-mould-ſhot	1
Jaundies	7
Impoſthume	8
Infants	22
Kingſevil	4
Lethargy	1
Livergrown	1
Meagrome	1
Palſie	1
Plague	4237
Purples	2
Quinſie	5
Rickets	23
Riſing of the Lights	18
Rupture	1
Scurvy	3
Shingles	1
Spotted Feaver	166
Stilborn	4
Stone	2
Stopping of the ſtomach	17
Strangury	3
Suddenly	2
Surfeit	74
Teeth	111
Thruſh	6
Tiſſick	9
Ulcer	1
Vomiting	10
Winde	4
Wormes	20

Chriſtned { Males — 90, Females — 81, In all — 171 } Buried { Males — 2777, Females — 2791, In all — 5568 } Plague — 4237

Increaſed in the Burials this Week — 249

Pariſhes clear of the Plague — 27 Pariſhes Infected — 103

The Aſſize of Bread ſet forth by Order of the Lord Maior and Court of Aldermen,
A penny Wheaten Loaf to contain Nine Ounces and a half, and three half-penny White Loaves the like weight.

Plate 4 A London Bill of Mortality for the week 15–22 August 1665

American underground test made in the Aleutian Islands in November 1971.

Besides fall-out there are radioactive materials associated with nuclear energy developments in Britain which, though well protected, are potentially highly dangerous. These include the reactors used to help generate electricity at Sizewell, Bradwell, Dungeness, Hinkley Point, Berkeley, Oldbury, Trawsfynydd, Wylfa, Hunterston, Dounreay, Chaplecross, and Calderhall. Reactors are always enclosed and fully protected within 'biological shields' and radioactive contamination is extremely unlikely. Yet accidents can happen, as in the case of the Windscale Plutonium Plant reactors in Cumberland which resulted in serious contamination of the local atmosphere with radioactive iodine-131. This, fortunately, has a relatively short half-life. Indeed, radioactive contamination is as yet small when evaluated in terms of natural background radiation; but the world is only on the threshold of the nuclear age and the potential danger is enormous.[9]

One further factor is worthy of consideration in the context of health and disease. Technological advances accompanying the machine age, such as motor cars, lifts, and automation generally, have made work increasingly sedentary in character and brought the doubtful benefits of increased leisure time. In theory such leisure time should be beneficial, but in fact it is fraught with social problems when it is employed passively rather than actively, and the reduced physical activity can have an unfavourable effect on physiological mechanisms. The way in which the weekend exodus to the seaside resorts or the countryside results in traffic jams is but one illustration. Evidently the manner in which leisure is used not infrequently involves mental strain, frustration, stress, and little genuine relaxation. In order to cope with the *onus* of lengthy leisure some individuals indulge in the use of hallucinogenic drugs, others undertake second remunerative jobs to fill their idle hours. Automation is bringing about important changes in the social environment and it is evident that increasing attention needs to be directed to the planning of facilities for exercise, recreation, entertainment, and hobbies for those unable to plan their own.

Cigarette smoking is a social habit of long standing. It has been praised variously, as a stimulant for work, as a sedative for relaxation, as a cure for colds and catarrh (so common in Britain), and indispensible for social intercourse. More accurately it must be damned as a major health hazard. It is a major aetiological factor in chronic bronchitis and ischaemic heart disease, and the increased incidence in lung cancer observed since the 1930s has resulted in cigarette smoking and atmospheric pollution being associated with that disease epidemiologically. Carcinogenic hydrocarbons, notably 3,4-benzpyrene among others, have been isolated from tars of cigarette smoke and from soot of polluted atmospheres. Why this relatively recent awareness of the association when seemingly the dangers were appreciated as early as the sixteenth century?

> Od's me, I marle what pleasure or felicity they have in taking this roguish tobacco. It's good for nothing but to choke a man, and fill him full of smoke and embers: there were four died out of one house last week with taking of it, and two more the bell went for yesternight; one of them, they say, will never scape it; he voided a bushel of soot yesterday, upward and downward. By the stocks, an there were no wiser men than I, I'd have it present whipping, man or woman, that should but deal with a tobacco pipe: why it will stifle them all in the end, as many as use it; it's little better than ratsbane or rosaker.
>
> Cob, in *Every Man in His Humour* by Ben Jonson

Royal disapprobation of tobacco—'the new drug from the Indies'—was offered also by King James I in *A Counter-blast to Tobacco* in 1604.

At the beginning of the twentieth century, 80 per cent of the tobacco consumed in the country was used in pipe tobaccos and only about 12 per cent in cigarettes. By 1914 cigarettes were fast catching up with pipe tobaccos and by the end of World War I, had overtaken them. Since then cigarettes have gone ahead rapidly and in the 1970s account for well over 80 per cent of the total consumption of tobacco in Britain. It takes twenty or thirty years before ill-effects become manifest and it was not until the 1930s, and the alarming rise of male deaths from lung cancer in the

1940s, that the full dangers were realised. It has been estimated that in Britain in the early 1970s one person dies from lung cancer every twenty-five minutes and that four times as many people every year die from lung cancer as are killed on the roads.

The prejudice against cigarette smoking by women was first broken in World War I, but it was not until the late 1920s and the 1930s that smoking by women started to become at all general in Britain. Now, women account for nearly one-third of the total consumption of cigarettes in the country and over 40 per cent of the female population aged 15 years and over are smokers. Regrettably the lung cancer mortality pattern for women is following that of the male population. Seemingly, pipe and cigar smoking produce a relatively small risk of lung cancer, possibly because pipe and cigar smokers seldom inhale the irritating alkaline smoke produced whereas cigarette smokers inhale the slightly acid smoke of cigarettes. The British smoker retains his cigarette until it has been smoked to a very small stub. The unburnt part of the cigarette acts as a filter and some of the smoke from the first half of the cigarette is condensed in the second half. When the second half is burnt the deposit within it is redistilled so that the smoke of the second half of the cigarette contains a higher concentration of harmful materials than the first. A change of habit to that of the American, who, because his cigarettes are cheap, can afford to throw away the cigarette when the stub is still relatively long, might reduce the risk.

Other human habits have associations with disease. Gluttony leads to obesity ('the Pickwick Syndrome') and accompanying ill-health, chronic alcoholism to mental deterioration in the form of impaired thinking and psychoses, and also to actual physical deterioration in the shape of gastritis and cirrhosis of the liver, and sexual behaviour in the permissive society to venereal disease. Britain, in common with several other 'developed' countries, appears to be caught up in an epidemic of drug dependence in the opiate, cocaine, amphetamines, and cannabis range; the long-term health effects of the latter two are impossible to assess. The taking of drugs seems to be particularly prevalent among those under the age of 30.

When the population of Britain was largely rural, people remained in the same locality for the greater part of their lives. Today people are mobile to a remarkable degree and there is a slackening of close family and social relationships. Many urban dwellers have moved further away from the centres of towns and their places of work and taken up residence in suburbs, in near-by market towns and villages, and in 'new towns' such as Crawley, Stevenage, Basildon, Corby, Peterlee, Cwmbran, East Kilbride, and Cumbernauld. In theory the 'new towns' were intended to reduce the journey to work, but there is still considerable mobility of workers. Such changes of place of residence involve not only a physical disturbance but also a social disturbance since people are obliged to create entirely new social environments for themselves. There is also the inevitable journey to work. This may be long or short, but it involves extra energy and takes toll of physical and mental reserves. What in theory may be a five-day week of 40 working hours may, because of the journey to work, amount in practice to 45–50 hours. Bus or rail travel to work involves waiting and standing; car driving often involves heavy traffic, blockage, and frustration. Agricultural communities and some working-class communities in the industrialised quarters of our towns which are not obliged to commute find certain endearing qualities in their environments. They enjoy a sense of 'belonging' and hand down traditions from father to son.

The increased mobility of the present population of Britain associated particularly with the motor car highlights the hazard of the road accident and the disease hazard of many of the lay-bys which are simply open air 'privies' used by thousands of people in need of a place of relief. At the lay-by there is nothing but the natural resources of the earth, and nature, though receptive, is not infinitely so. Many such places have become bywords for filth and are a potential source of disease.

The high degree of internal mobility which characterises the population of Britain in the latter half of the twentieth century is matched by ever-increasing movement of people between this country and other parts of the world. Movement of people in association with trade has existed since time immemorial. Men, goods, and animals have followed the main trade routes of the world; so, too, have the germs of disease (*see* p 93 (plague), p 167 (cholera) etc). Contributory factors, involving human contacts, have included commercial exchanges, caravans, pilgrimages, movements of labour and migrations in search of pastures and water for livestock. Germs have passed from man to man, from animal to animal, or from animal to man, or indirectly through carriers. Man's lines of communication have long been the pathways of infection.

India, for instance, suffering from conditions of extreme overcrowding, undernourishment, lack of hygiene and grinding poverty, has endemic cholera, plague, malaria, typhoid, and dysentery. China is an endemic centre for leprosy and has been the cradle of two plagues, bubonic and pneumonic. South America is the home of yellow fever. Communications between these disease foci and Europe and Britain in the past were by sea, and the journey to Britain was usually longer than the incubation period for most infectious diseases. This was an effective protection for Britain against infection. However the increase in the volume and speed of modern air travel has broken down this protection and exposed Britain to the danger of direct transmissions of disease. The last decade or so has seen an appreciable movement of Indians, Pakistanis, and West Indians into the country, drawn particularly to London, Birmingham, and some of the major cities. Not infrequently these people have presented serious health risks in that some are carriers of tuberculosis, typhoid, and other infectious diseases.

The discovery of the effectiveness of antibiotics, particularly chloramphenicol and the synthetic penicillins, has almost eliminated the risk of death from bacterial infectious diseases, but it may be wondered whether medicine's present advantage over infection is going to last. Some drugs themselves are often poisonous—penicillin can kill people who have become sensitive to it. Some drugs often work too well. They can eliminate other organisms that are, in fact, protecting the body. Treatment with broad-spectrum antibiotics often produces an 'overgrowth' of fungal infection of the mouth, skin, and genitals, usually only irritating, but on occasion, serious. The constant indiscriminate use of drugs inevitably produces drug-resistant strains like *Pseudomonas*, a hospital-bred organism which, when established, has an 80 per cent mortality rate. Resistant strains are countered in turn by new drugs, but it means that the pharmaceutical industry is hard pressed simply to maintain the *status quo*.

The rapidly changing social environment associated with Britain's twentieth-century technological civilisation is bringing a whole host of new and unsolved health and allied problems in its wake.

Notes to this chapter are on pp 248–9.

6

Pre-Norman and Norman Times

I

Before considering the diseases thought likely to have been prevalent in the Norman and pre-Norman period it is necessary to examine briefly the environment of Britain's earliest people. The landscape provided the physical stage on which the drama of British history was to be performed. The actors were the prehistoric, Anglo-Saxon, and Scandinavian populations joined during the eleventh century, by the Normans.

Living in a highly industrialised and urbanised society and occupying a countryside which has been very much modified by the hand of man, it is difficult to visualise what the scene was like in Britain in those early days. What is certain is that there are few, if any, places in the country which look today as they did in pre-Norman times. Changes during the last fifteen centuries have been such that the appearance of the countryside then must have been so different as to be quite unrecognisable to present-day inhabitants of these islands.

Pine forests established themselves throughout the greater part of Britain during the milder conditions which followed the retreat of the ice of the last glacial epoch. The lowlands of Britain were later covered with heavy forests of oak, elm, beech, and ash, and the uplands with pine and birch up to the heights of 1,500–2,000 feet. Limestone ridges and chalk lands carried only a light forest cover. The marsh, bogs, and fens of the lowlands carried scrub. Beyond the forested heights came the open moorlands covered with expanses of tundra-type vegetation.

Britain's virgin or primitive landscape was, to say the least, inhospitable to early man and yet, as already noted, the country was destined to become a final landfall for several migrating European peoples.

Following the physical separation of Britain from the Continent successive groups, including traders in copper, tin, and gold and husbandmen in search of fresh pastures and soils, reached what are now Cornwall, Pembrokeshire, Anglesey, and the west coast of Scotland from the Mediterranean. Others, of different stock, took the shorter crossing and entered Britain at its south-east corner. The newcomers brought knowledge of agriculture, of domestic animals and grain, of the hoe, the grinding stone, of weaving clothes and fashioning clay pots. They were small, dark, slender Mediterranean or Iberian colonists, few and scattered at first, who tended to group around the western shores of the country. Their habitations were stone or wooden huts covered with branches and surrounded by small fields hewn or burnt out of woodland.

In the fullness of time came the Celts or Gaels, tall, blue-eyed folk, who crossed Europe from the east and settled in Gaul or what is now France. They came first as small bands and families, later in tribal armies. Early Celts lived in hill-forts occupying strategic and defensive sites. Their forts might not have been much more than an enclosure surrounded by a single ditch and bank; at other times the protective works might have comprised multiple lines of earth-works and dry stone walling. The Celts used iron tools and created permanent fields and villages. They used small wooden ploughs drawn by oxen to till light chalk soils for crops of wheat, barley, and oats. Celtic peoples who came later into south-east Britain introduced a heavier plough which, with coulter and mould board, was capable of undercutting and turning over the sod of the heavy clay lands. Increasing control over the environment through the use of iron for tools enabled the settlers to utilise land which was open or thinly forested for flocks, herds, and crops and attack 'damp' oakwood forests which covered the low-lying clay lands. Prehistoric contributions to the modification of the primitive landscape of Britain were not very great, though developments in agriculture and the appearance of settled villages are thought by some to have represented a 'revolution in human life comparable in magnitude with the effects of the industrial revolution of the nineteenth century'.[1]

The population of the country increased only gradually during the Celtic occupation. It totalled little more than a quarter million

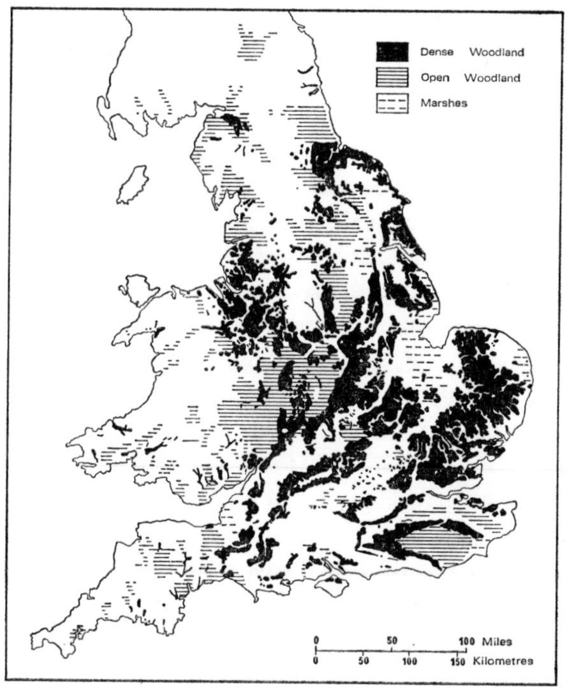

Fig 30 Woodland and marsh in Britain in Roman times (*based on a map of Roman Britain published by the Ordnance Survey*)

and was unevenly distributed. Indeed, no more than a fifth of Britain was occupied and this mainly in the south-east.

During their stay in Britain the Romans left a visible imprint on the landscape. They introduced towns, agricultural 'villas' and a network of roads and yet, relative to the wilderness that was Britain (Fig 30), such man-made features of the landscape were minimal. Roman Britain had three culture zones. First, the Civil Zone of southern and eastern England, with all the signs of a peaceful occupation. Second, the Military Zone comprising Wales, northern England, and southern Scotland between Hadrian's Wall and the Antonine Wall,[2] where the military occupation was maintained by forts. In Scotland, north of the Antonine Wall was a third non-Roman zone. Here the Roman occupation was short-lived and the indigenous peoples retained their traditional habits and customs.

Municipal life was a feature of Roman civilisation and at the height of the Roman occupation there were about fifty towns in southern Britain. The largest London (Londinium), which replaced Colchester as the leading mercantile centre of the country, may have had 20,000 inhabitants. By modern standards the other towns were small, with no more than 2,000 to 5,000 inhabitants. St Albans, Colchester, Lincoln, York, Gloucester, Caerleon, Chester, and Bath may have averaged 3,500 each. There were about a dozen other towns with 2,000 or so inhabitants each and forty to fifty other places with about 1,000 each. Fleure[3] has suggested a total urban population of 120,000 and says 'if it was about 20 per cent of the whole, this gives a total of 600,000 for prosperous times in the areas south of Hadrian's Wall'. North of Hadrian's Wall the figure would be a mere fraction of this amount.

The streets of the Roman towns were broad and well-paved. They ran straight and parallel and crossed at right angles. Houses were usually of the single-storey type, each detached and standing in a garden. The Romans excelled in their stone buildings. They were solid, well designed, and ventilated. A unique feature was the hypocaust or heating chamber built beneath the floors from which warm air was distributed by means of a system of pottery flue pipes into the walls of the main living room.

The Romans built public water supplies and paid attention to the purity of the water. At certain points in the aqueducts were built settling basins in which suspended solid matter was able to sediment out. The Romans were presumably not aware of the fact, but such storage of water in a relatively quiet state was effective in ridding it of some of its harmful bacteria. There were also bath-houses and sewers for the prompt disposal of excrement.

In fact there were latrines in Hadrian's Wall (eg at Housesteads), flushed by running water from storage tanks or surface water. The realisation of the importance of a clean and ample water supply and the prompt disposal of polluted water after use must have been of incalculable value in terms of public health. It meant that water-borne diseases such as typhoid fever were reduced and typhus and relapsing fever curbed through personal cleanliness and the accompanying discouragement of lice.

Scattered within the countryside of the Civil Zone were farmsteads or villas, worked by Romanised Britons, and on which wheat and barley were grown. These substantial structures contrasted with the foul and verminous conditions of the rude thatched shelters with straw-covered floors and smoke-laden atmospheres which housed the ancient Britons.

The Romans introduced poultry, pheasants, pears, cherries, figs, mulberries, and a wide variety of herbs into Britain, together with such vegetables as cabbage, onions, turnips, lettuce, parsley, and parsnips. Nutritionally they had a more balanced and healthy diet than the native British population.

The departure of the Roman legions was followed by a period of disorder among the Britons. This encouraged foreign invasion. Angles, Saxons, and Jutes from northern Germany and Denmark invaded England by way of the Humber, Wash, Thames, and Solent. They came first as raiders but later as settlers, and, by AD 600, had conquered all but some western and northern areas. Celtic tribes held out in Cornwall, Wales, Strathclyde and other parts of Scotland but elsewhere Romano-British inhabitants were submerged under the Anglo-Saxon flood. The Angles, Saxons, and Jutes were illiterate, heathen peoples. They destroyed towns and villages, overthrew the Christian Church and, by displacing the Latin language, suppressed the art of writing and also the men who could practise it. There is in consequence such a lack of written records for this period that the title Dark Ages seems appropriate.

The natural harbours of the east coast provided easy landfalls for the invading Anglo-Saxons who occupied sites along river banks or on the lower slopes of valleys. These land-hungry people evidently recognised the good ploughlands and waterside flood plains which would make the best grazings. Indeed the river systems provided the key to much of the distribution of settlement in Anglo-Saxon times. Fox[4] has suggested that 'it is archaeologically possible to see the Saxon farmer at work, turning the valley bottoms into water meadows, the forest margins into arable and pasture'. Evidently they came as settlers determined to till the land.

Anglo-Saxon and Scandinavian settlement spread over some twenty generations prior to AD 1066. During that time there was a valleyward movement of people consequent upon the progressive clearing of woods and draining of marsh. Indeed within Lowland Britain there was a slow change over from what was dominance of the environment to dominance of man. But the task was immense. Forests were dense and many and the numbers of people engaged in clearing very small. Roads were few, mainly ridgeways along the downs and those Roman roads which were still usable. In consequence there were few if any facilities for communications and each district was more or less dependent on its own resources (Fig 31).

In the west and north of the country Celtic tribes maintained a traditional way of life. They grazed their cattle and sheep on bleak open moorlands in summertime but their permanent homes were small farms or communal enclosures on the open slopes of valleys or the brows of hills above the tangle of forest and undergrowth in the valley bottoms and on the coasts. Christianity persisted among them at a time when Saxon heathendom prevailed in the south and east of the country. Contact with Rome and Christian settlements in the Mediterranean was maintained along western sea-ways linking Wales, Ireland, and south-west England with Brittany and Spain. Christian teachers—the Celtic saints—established a large number of semi-monastic settlements and showed other less venturesome people how the forest lands might be tamed and planted with grains or reclaimed as pasture. Saints, such as St Columba, St Ninian, St Samson, and St David, were among the pioneers of the valleyward movement and village settlement in the west and north of Britain. Estimates of the total population of Britain at this time are necessarily guesswork, but with the reservations that befit guesswork it would seem that there were still less than a million people in the country.

Viking invasions in the eighth, ninth, and tenth centuries resulted in the seizure of not only East Anglia but lands to the south in what is now Essex, Hertfordshire, Bedfordshire, and

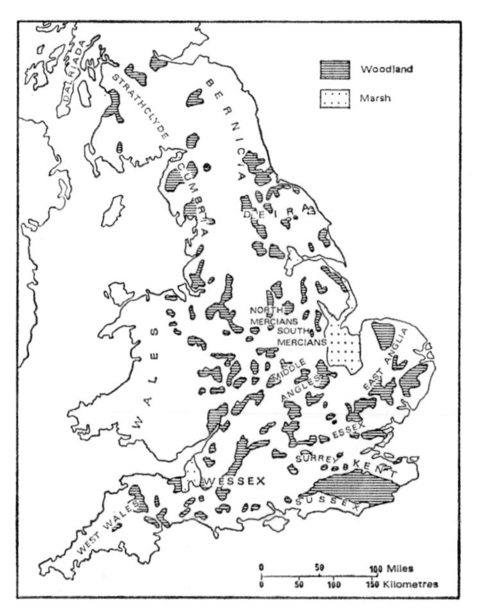

Fig 31 Woodland and marsh in Britain in the late seventh century (*based on Hodgkin 1959*)

possibly Buckinghamshire. Danes also held lands in Yorkshire, Lincolnshire, Leicestershire, Nottinghamshire, Northamptonshire, and Derbyshire. Other Viking raids, invasion, and colonisation of the Orkneys, the north of Scotland and the Hebrides heralded a later expansion into Ireland. From there and from their kingdom on the Isle of Man they invaded England and the Scottish Lowlands (Fig 3). These activities along the northern and western seaways coincided with a general improvement of climate which was, in general terms, dry and warm. This particular climatic amelioration reached its optimum probably between AD 800 and 1000.

The last successful invasion of Britain was by the Normans in AD 1066. However, their arrival was, in effect, little more than the transposition of an aristocracy to Britain and not a folk movement of people in search of a new homeland. The Normans merely reinforced and partly replaced by their own feudal system an aristocracy which had already been established. The Welsh coastlands yielded slowly to the Norman advance but northern England remained a frontier province as during the Dark Ages. Mainland Scotland was nominally united under one King.

The great achievement of the Anglo-Saxons and the Vikings was their progressive clearing of the natural woodland, though a great deal still remained even in Norman times, as the Domesday Survey (1086) bears witness. There were also lands converted to 'forest law' to preserve the King's hunting, and much devastated land following the Norman Conquest, particularly in Yorkshire. Even so, arable land continued to increase at the expense of woodland throughout England and in the Celtic lands of the north and west. There were inroads, too, on the marshlands of the Thames, Fens, Somerset Levels, Humber Lowlands, Holderness, the mosslands of Lancashire and Central Scotland (especially the Vale of Menteith and Carse of Gowrie) and other local improvements following the drainage of floodable valleys.

Population distribution continued to show a fairly close relationship to agricultural productivity, the greater part being located in Lowland Britain, especially south of a line from the Wash to the middle Severn (Fig 32). There were, however, wide variations in the density of the population, since thousands of square miles remained forested or untouched by the plough. The average density of population in England was about 15 to 20 per square mile although parts of East Anglia might have had as many as 20 to 40 per square mile. It is doubtful if the average was more than four people per square mile over the whole of northern England. In Wales and Scotland it was probably less. Hoskins[5] says that Norfolk (95,000), Lincoln (90,000), Suffolk and Devon

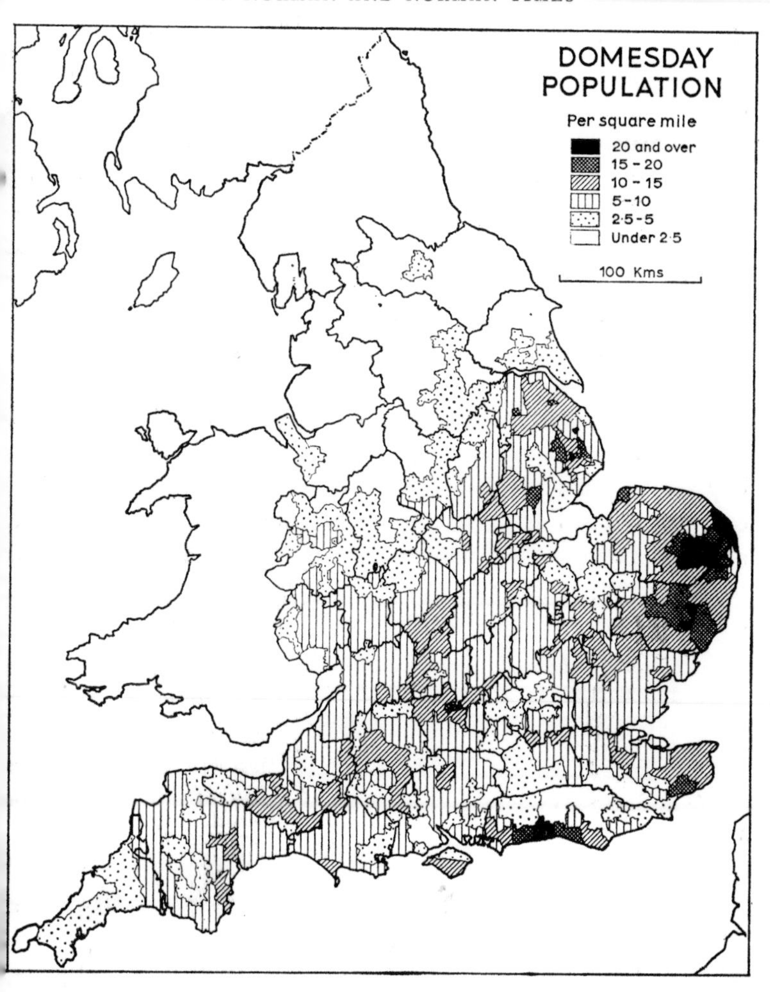

Fig 32 Distribution of population in England at the time of Domesday (*after H. C. Darby 1970*)

(70,000 each) were the most populous counties, followed by Kent, Hampshire, Sussex, and Wiltshire, each with 40,000–50,000 people. What is certain is that the country was still very sparsely populated. The population for the whole country at the time of the Domesday Survey was little more than 1½–2 millions. In England small nucleated villages had evolved within the forest clearings and 'towns' appeared, or reappeared, in rudimentary form. In the north and west of the country, however, the characteristic settlement pattern was that of the scattered hamlet or single farmstead surrounded by a few small fields.

Urban life had been introduced into Britain by the Romans, but there was virtually no continuity between the towns of Roman Britain and those of later times. Indeed the very sites of some Roman towns were deserted during the tempest of invasion. Following the firm establishment of the Anglo-Saxons and later Scandinavian settlers there once more came into being centres with an economy different to that of the surrounding agricultural country side. Some of these grew up as market towns serving as foci for the villages around and those favoured by location became regional capitals for wider areas. Others, along the coast, developed as protective burghs against Danish invasion, and later became ports with overseas connections. A number of towns originated as seats of administration for King, Army, or Church (Exeter, Chichester, Norwich). Such were some of the Midland boroughs, each surrounded by a shire to which it gave its name, eg Bedford, Leicester, Nottingham. Several existing settlements gained borough status by charter.

Society, based on the manor or barony, was feudal. Nine-tenths of the peasant cultivators were tied to the land and lived in small villages. Elsewhere they occupied isolated farmsteads.

Words such as beef (*boeuf*), mutton (*mouton*), and pork (*porc*) were introduced into the English language by the Normans but these items of food were characteristic of the meals of nobles rather than of peasants. Ox, sheep, swine, were the names used by the herdsmen who tended the animals rather than fed upon them. Feeding farm animals presented serious problems with the approach of winter. There was virtually no fodder, so it was customary to slaughter old and weakly beasts in the late autumn.

2

The amount of reliable material relating to disease in Britain up to and including Norman times is lamentably small, and when information is available, it lacks detail. During the Roman occupation it is assumed that their society included physicians, surgeons, oculists and others to care for the health of the population but few traces of their existence and activity remain. Very little is known about the diseases of the time. The Dark Ages are darker to the student of disease than to the historian. What little information there is, is based largely on palaeopathological material,[6] on state documents, and, since literacy was associated with monastic clergy, on documents which survived the Viking invasions or the destruction and disposal of monastic libraries in the time of Henry VIII.

Because they leave few traces on early human skeletons, some of the diseases of prehistoric times are difficult to study. Other diseases, however, leave bony manifestations which permit of diagnosis. Palaeopathologists including Brothwell, Møller-Christensen, Calvin-Wells, Sandison and McArthur, have studied between them thousands of early British skeletons and, on the basis of inferential evidence indicate the presence of leprosy, syphilis, tuberculosis, osteo-arthritis, tumours and dental caries in the pre-Conquest population of this country.

It is not possible to indicate the frequency of leprosy from the contemporary historical records because available descriptions do not provide convincing evidence that what was called 'leprosy' was in fact produced by the *Mycobacterium leprae*. The medieval conception of leprosy was strongly influenced by the biblical usage of the word. In the Bible 'leprosy' (*zara'ath*) is more a generic word embracing a number of different diseases. It also implied 'moral uncleanliness', which explains the present-day overtones of the word leper. MacArthur[7] sums up the confusion in diagnosis in early records as follows:

> In the past, 'leprosy' and its equivalents had a multitude of meanings. It was used for the true disease and for every disorder that was formerly supposed to be leprosy. The Greek form of the word was *lepra* (*lepros*, scaly), and was applied by the Greeks themselves to scaling skin diseases of the psoriasis type, and never to leprosy for which they used the word 'elephantiasis' because of the thickening and corrugation of the skin. Unfortunately, 'lepra' was adopted as the classical medical term for leprosy with the result that, by

suggestion of the word itself, a host of skin conditions associated with scales or scabs, which have no connexion with real leprosy, were identified as manifestations of this disease.

Leprosy began to make its appearance in Europe in the sixth century and it seems likely that the disease was introduced into Britain later in the same century. The lazar house (*leprosarium*) on the Island of Tean in the Isles of Scilly has been dated to the seventh century. Doubtless the so-called lazar houses included among the inmates genuine cases of leprosy but the word 'lazar' signifies 'a poor and diseased person, especially a leper' and derives from the Lazarus of the Gospels. MacArthur says that there is no scriptural record 'that either the symbolic beggar of Christ's parable or the real Lazarus of Bethany, suffered from leprosy. . . . The beggar was *ulcerosus*, full of sores, which then suggested leprosy only'. Lazar houses, therefore, such as, for example, the early Hospital of St Peter and St Leonard at York, founded in AD 936 by King Athelstane,[8] were not necessarily built to halt the spread of the disease.

During the winter months the lack of fresh foods, meat, fruit, and vegetables might be expected to have caused a serious deficiency of vitamins A and C. This would give rise to increased liability to septic infections of the skin, to rough and dry skin, or to mild scurvy. This is inferential. There is no direct bone evidence. Famines were frequent and it has been suggested that the skin disease might have been pellagra, a dietary deficiency disease, and not leprosy. Either way personal hygiene during Anglo-Saxon times, and indeed until the mid-nineteenth century, was of a low order (where, except in literature, does one find evidence for unpleasant, but medically important environmental facts such as the ubiquity of infestation by human fleas?) and skin diseases were commonplace. There is skeletal evidence of nine cases of leprosy in Britain, the oldest dating back to about AD 600. In Scotland, St Fillan the 'leper', a teacher of the sixth century of peculiar sanctity, was specially celebrated in the cure of the disease (plate p 65).[9]

Investigators into the origin of syphilis have become more cautious in expressing opinions about evidence for the disease in ancient times. Syphilis has to be studied in terms of the treponematoses as a whole, since venereal syphilis, yaws, and endemic syphilis cause bone lesions. At present there is no way to differentiate between these three infections in bones if, bearing in mind the importance of bacterial mutation, they are to be regarded as individual infections by separate organisms. The sole British example in which the diagnosis of advanced syphilis is unquestionable is of a female skull discovered in the graveyard of St Mary Spittle, Spitalfields Market, London, in 1926 (plate p 48). There are two skeletons, one from Chadlington, Oxfordshire, and another from near Portsmouth which provide evidence that tuberculosis was also established in Britain by Saxon times.

As with modern man, the earlier peoples of Britain suffered from several varieties of arthritis. The type generally known as osteo-arthritis was seemingly common.

Apart from ancient bones there are early literary sources which afford evidence for the presence of certain diseases in Norman and pre-Norman times. Creighton,[10] the nineteenth-century medical historian, tells of a foreign invasion of plague at the time of the Venerable Bede (AD 672–735), of famine, pestilence, and of some non-famine sicknesses, but there is insufficient information to provide even tentative support for the diagnosis of any of these diseases.

Plague has always been a major scourge of mankind but in these early days 'plague' was a general word for all diseases with a high mortality. It was in fact by way of being a generic term, rather like 'leprosy' was, or 'influenza' or 'fever' are today. The name might have included true plague (bubonic or pneumonic) yet might also have indicated typhus or some other disorder for which no specific name existed. The word 'pestilence', too, would refer to almost any kind of acute epidemic. Pestilences were so frequent that only the most virulent and fatal ones would have been recorded. Nevertheless they were important factors in the social history of early Britain. Medical opinion at the present time would say that people do not die of hunger *per se*, ie hunger associated with the so-called 'famine pestilence', but rather from disease contracted as a result of under-nourishment.

A major difficulty in attempts to interpret medieval and indeed classical medicine from literary documentary evidence is the complete lack of modern terminology and nomenclature. For instance, the Anglo-Saxon word for disease was *morbus*, that for pain was *ádl*, *côau*, or *ece*, and for ache or pain, *waerc*. When the ache or pain was localised in a particular part of the body, or was especially severe or fatal, these words were compounded with others, eg *fót-ádl*, a pain in the foot, gout; *ban-ĉoau*, a killing pain, erysipelas. The usual word for pestilence was *máncwealm* (*cwelan*, to die), for cattle murrain *orfcwealm* and both pestilence and murrain were signified by *wól*. It is clear that there is little evidence to suggest specific knowledge of the interior organs of the body or of their diseases.

In accordance with the beliefs of the times plagues and pestilences were attributed to a variety of causes including arrows shot at victims by the gods, failure of crops, movement of stars, storms, the effect of drought or floods. In Anglo-Saxon England the sudden onset of disease was often ascribed to 'elfshot', arrows shot by elves. An alternative theory was that pestilence was caused by corrupted air, emanations from marshes or noxious vapours inhaled into the body. St Columba on Iona, seeing a dense cloud arising from the sea on an otherwise clear day, said to one of his monks:

> This cloud will be very harmful to men and to cattle . . . it will pour down in the evening a pestilential rain which will cause grievous and festering ulcers to be found on the bodies of men and on the teats of cattle; and by these the sick men and cattle will suffer from that poisonous infection even unto death.

Such views persisted up to the beginning of scientific study of medicine, barely a century ago.

Pestilence appears to have devastated parts of Ireland, Scotland, and Wales in the mid-sixth century but there is no record of it having spread into England. St Brioc is said to have returned to his native Ceredigion (Cardiganshire) from Brittany to minister to his folk during a local outbreak of pestilence about AD 526. In AD 550 'the Yellow Plague was roaming through the land in the guise of a loathly monster'. A full description of this outbreak of pestilence in Wales is given in the life of St Teilo in the *Liber Landavensis* (quoted by Bonser[11]).

> St Teilo received the pastoral care of the Church of Llandaff . . . in which however he could not long remain, on account of the pestilence which nearly destroyed the whole nation. It was called Pestis Flava, because it occasioned all persons who were seized by it, to be yellow and without blood, and it appeared to men as a column of watery cloud, having one end trailing along the ground, and on the other above, proceeding in the air, and passing through the whole country like a shower going through the bottom of the valleys. Whatever living creatures it touched with its pestiferous blast, either immediately died, or sickened for death. If any one endeavoured to apply a remedy to the sick person, not only had the medicine no effect, but the dreadful disorder brought the physician, together with the sick person, to death. For it seized Maelgwn, King of Guenedoth, and destroyed his country; and so greatly did the aforesaid destruction rage throughout the nation, that it caused the country to be nearly deserted.

It is important to view such literary references within their traditional framework since pestilence at that time was considered to be a living thing which roamed the land. There are several references in Celtic sources to severe epidemics in Ireland in the period AD 537–77. Mention is made of 'the battle of Camlann, in which Arthur and Medraut fell: and there was a plague (*mortalitas*) in Britain and Ireland', of 'a great mortality in which Mailcun, King of Guededota, reposed', of 'the disease (*pestis*) which is called *samthrosc*', of 'the Great Pestilence called the Boy Connell (buidhe chonnaill) began' (in AD 550), and for the year AD 555 'a great mortality in this year'.

Scotland was also visited, for Adamnan,[12] in his *Life of St Columba*, is at pains to record that the pestilence did not penetrate to those areas of Scotland in which were situated monasteries founded by St Columba. Shrewsbury,[13] is of the opinion that these epidemics were spread by Celtic missionaries who carried them from the Continent to Ireland, whence they spread to Wales and Scotland.

The Celtic names for the pestilence, *blefed*, *cron chonnaill*, and *buidhe chonnaill*, have given rise to misunderstanding as to the nature of the diseases concerned. Shrewsbury identifies all three words as smallpox, whereas MacArthur[14] suggests bubonic plague, relapsing fever, and smallpox respectively. MacArthur is of the opinion that the series of outbreaks of pestilence (*blefed*) in the sixth century, were bubonic plague and related to the pestilence in the Byzantine Empire in Justinian's reign, the first authenticated visitation of bubonic plague in Europe. This epidemic probably originated in the hinterland of south-west Asia about AD 540. Pelusium,[15] the great commercial entrepôt of Egypt, served as the major centre from which it is thought the infection spread, probably by corn ships. By AD 542, the pandemic extension of the great pestilence was under way and it eventually spread, slowly, throughout the then known world, dying out towards the end of the sixth century. Western Europe, especially southern France and Germany, was heavily smitten by the *Lues inguinaria* (presumably the buboes of bubonic plague) with far-reaching social and economic results. The view of Procopius,[16] an eye-witness, was that 'no cause for the plague could be given or imagined except God'.

A second series of outbreaks of pestilence began in Britain and Ireland in AD 664 and continued, with short intervals, for about half a century. Creighton refers to this invasion as the only epidemic in early British annals that could be regarded as a plague of the same nature and on the same scale as the devastation of the continent of Europe brought by the plague of Justinian a century earlier. Community life was disorganised and Bede[17] speaks of the pestilence depopulating the south coast of Britain before spreading north into Northumbria. As with most visitations this pestilence was reputedly heralded by natural phenomena, in this case an eclipse of the sun, and that 'it ravaged the country far and wide and destroyed a great multitude of men'. This was the great plague of Cadwallader's time. The East Saxons turned to idolatry on account of it. 'While the plague caused a heavy death roll in the province', Sighere, who was ruler of the East Saxons under Wulfhere, King of Mercia, and his people 'abandoned the mysteries of the Christian Faith and relapsed into paganism. For the King himself, together with many of the nobles and common folk, loved this life and sought no other, having no belief in a future life'.[18]

The epidemic disease raged with equal severity in Ireland and Scotland. St Cuthbert was seized by the pestilence in the Abbey of Old Melrose. Seemingly he developed the usual swelling in the groin which unfortunately burst inwards so that he suffered from its effects for the rest of his life. Boisil, the prior of the Abbey, succumbed to it.[19]

Pestilence is recorded in Britain and Ireland in the years following AD 664, particularly in the monasteries. At this time, before the growth of towns, monasteries were among the larger communities; they carried large stores of grain which would attract rats, the carriers of the plague. It is likely that monks, travelling from infested to uninfested monasteries, spread the pestilence. Certainly monasteries in isolated and unlikely places were afflicted. Bede tells of a decree by Theodore at the Synod of Hertford in AD 672, forbidding the movement of monks from one monastery to another 'after the Celtic fashion'. Outbreaks in Wales in the early 680s are mentioned in the Chronicle of the Princes (*Brut y Twysogion*) and for Ireland in the Cambrian Annals (*Annales Cambriae*).

There was a series of outbreaks of pestilence in the second half of the eighth century, chronicled in Irish annals, and from the pages of the *Anglo-Saxon Chronicle* come a long record of famine and disease in the ninth, tenth, and eleventh centuries. These coincided with the incursions of the Norsemen in northern England and the Danish invasions of East Anglia which ravaged the country, leaving destruction in their train.

In the absence of efficient means of communication areas devastated by Viking raiders were unable to obtain even the necessities of life from other districts and famines were frequent. There followed inevitable and local serious mortality among the population and murrain (not plague) among cattle. Creighton has listed 'famine pestilences' in England for the years 679 to 1322.[20] There were forty-two such pestilences during this period, some

Table 6 Famine pestilences in England between 1069 and 1143 (*after Creighton*)

Year	*Character*	*Authority*
1069	Wasting of Yorkshire	Simeon of Durham, ii, 188
1086 1087	Great fever pestilence Sharp fever	*Anglo-Saxon Chronicle*, Malmesbury Henry of Huntingdon, and most annalists
1091	Siege of Durham by the Scots	Simeon of Durham, ii, 339
1093 1095 1096 1097	Floods; hard winter; severe famines; universal sickness and mortality	*Anglo-Saxon Chronicle*. Annals of Winchester. William of Malmesbury. Henry of Huntingdon. Annals of Margan, Mathew Paris, and others.
1103 1104 1105	General pestilence and murrain	*Anglo-Saxon Chronicle*. Roger of Wendover
1110 1111	Famine	*Anglo-Saxon Chronicle*. Roger of Wendover
1112	'Destructive pestilence'	*Anglo-Saxon Chronicle*. Annals of Osney. *Annales Cambriae*
1114	Famine in Ireland; flight or death of people	Annals of Margan
1125	Most dire famine in all England; pestilence and murrain	*Anglo-Saxon Chronicle*. William of Malmesbury, *Gest. Pont.*, p 442. Henry of Huntingdon. Annals of Margan. Roger of Howden
1137 1140	Famine from Civil War; mortality	*Anglo-Saxon Chronicle*. Annals of Winchester. Henry of Hungtingdon (1138)
1143	Famine and mortality	*Gesta Stephani*, p 98. William of Newburgh. Henry of Hintingdon

lasting a year, others as many as four. An extract from the list for the second half of the eleventh and the first half of the twelfth century is given in Table 6.

There was virtually no increase of population during this period, famines and pestilences being the principal checks which kept it down to the level of its scanty means of subsistence. Expectation of life was probably less than thirty years.

Apart from pestilences the early Anglo-Saxons were familiar with 'spring relapses', a peculiarity of the malaria that flourished in parts of England at that time. In infected persons the parasites tended to lie dormant during the winter, later becoming active with the approach of the longer days in spring. The Fen country, the marshes of the Thames, and the marshes of south-east Kent were endemic centres for malaria, the *lencten ádl* (spring ill). MacArthur[21] says that the Venerable Bede in his *Ecclesiastical History*, written in Latin, describes the miraculous cure of a long-continued illness which he calls merely by the general term *febris* (fever). It is of interest to note that in the Anglo-Saxon translation of this work, attributed to King Alfred, the translator identified the malady as malaria, and, instead of employing one of the Anglo-Saxon equivalents of 'fever', he boldly replaced Bede's vague word by the specific '*Lencten ádl*', malaria. The term 'ague' has been used as synonomous with malaria, but not all agues were malarial. Ague originally meant any acute fever and later was often applied specifically to typhus, sometimes in the more unequivocal form 'the spotted ague'.

Creighton, an important source of information for this period, also mentions a swift and fatal pestilence which broke out among the Danes in Kent. It was probably of the same form of camp sickness, including dysentery (as the name *dolor viscerum* suggests) as that which occurred in later periods. It is the only instance of the kind recorded in early British history. It occurred in the year AD 1010 or 1011, when the Danes stormed Canterbury, massacred its inhabitants and carried the remnants captive to their ships at Sandwich. If the disease was dysentery it is now known that the organisms which produce it are widespread in those places where hygiene or sanitation are of a low order. Such conditions are favourable to the spread of the disease which produces a steady and unrelenting series of infections. Mortality from dysentery depends on the particular variety of, or organisms involved, but death rates were usually high. Creighton, quoting an early narrative of William of Malmesbury, says 'a deadly sickness broke out among the Danes, affecting them in troops (*catervatim*) and proving so rapid in its effects that death ensured before they could feel pain. The stench of the unburied bodies so infected the air as to bring a plague upon those of them who had remained well.'[22] Seemingly the pestilence destroyed the Danes by tens and twenties and a large number perished.

There are scattered references to 'pock disease' and 'the pox' in

Norman and pre-Norman times, and these would refer to small-pox, a disease which did not come into prominence until the sixteenth century and not to 'great pox' or syphilis, which erupted with extraordinary virulence in 1493 and spread with great rapidity. This soon reached Britain by 1497 and there were victims as far north as Aberdeen.

Though the treatment of disease is not considered in this book brief mention should be made at this point to the Anglo-Saxon Leech Books, and more precisely to the Leech Book of Bald. Bald, a physician, had a monastic scribe write, between AD 900 and 950, a medical manuscript which incorporated extensive borrowings from Greek writers such as Paulus of Aegina, Alexander of Tralles, and others of classical medical thought, together with a medley of folk remedies, Christian prayers, magic and pagan material. The Leech Book embodies some of the best medical literature available to Britain and western Europe at that time and also provides an indication of the separation between spiritual and physical conceptions of disease healing. The work is of interest in that it contains characteristic views on medicine and disease in these early centuries.

Notes to this chapter are on pp 249–50.

7
Medieval Times

The invasion of Britain by bubonic plague in 1348 was probably the chief medical event in medieval times. Apocalyptic estimates of plague mortality of one-third or even one-half of the population have long prevailed. 'The Great Pestilence' or 'The Great Mortality' as the disease was called by contemporary chroniclers (the modern pseudonym 'The Black Death' was not introduced until 1823) was a wave of the second pandemic of the disease which had spread out from the Indian subcontinent between 1340 and 1352 to affect Asia Minor, Europe, the Channel Isles, Britain, Iceland and Greenland, and parts of North Africa.

In keeping with the general beliefs of those early days, it comes as no surprise to discover that putrefying corpses in China, fogs over Europe, corruption or pollution of the air by noxious vapours, comet-borne miasmas, Jews acting as agents for Satan, and punishment by God for human sins, were among the several explanations invoked as the cause of plague.

The disease is thought to have started its expansion into Europe at Kaffa (Caffa)—the modern Feodosia—a Black Sea town on the Crimea, in 1346 (Fig 33). The town was under siege by the Tatar army of Kipchik Khan Janibeg and 'the beleaguered Christians saw the heavenly arrows strike the Tatars'. Seemingly the Tatars were suffering from plague which they had probably contracted in Central Asia. Gabriel de Mussis says that the Tatars hurled their dead with catapults into the city. This would suggest a fourteenth-century attempt at bacteriological warfare. Indeed, subsequent events revealed 'the Great Pestilence' to be a particularly vicious attempt on the part of plague parasites to wipe out the human race.

Bubonic plague was primarily an infectious disease of certain species of rodents and plague in man was the result of a fortuitous invasion of the human body by an internal pathogen of these

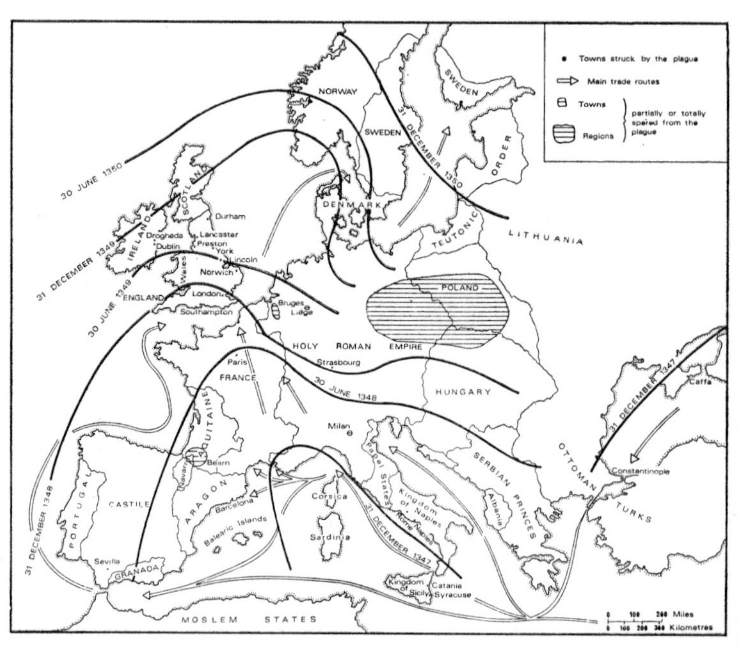

Fig 33 Advance of the Great Pestilence ('Black Death') into west and central Europe, 1347–50 (*adapted from Carpentier 1962*)

rodents. The pathogen in question was the bacillus *Pasteurella pestis* and this produced endemic infection in rats. It was transferred from plague-infested rats to man through the bite of infected rat fleas (*Xenopsylla cheopis*). The rodent acted as host to the plague bacillus and the flea was the carrier or vector. In medieval times the black or house rat (*Rattus rattus*) was responsible for the epidemics of bubonic plague. It infested the dark, unventilated, humid, wattle-and-daub and thatch-roofed dwellings of the common folk and rarely and reluctantly strayed out of doors. A native of Western Asia, this type of rat was a climbing animal which lived and bred in close contact with man. The field rat (*Rattus norvegicus*), which did not come to Europe till the eighteenth century but which now holds supremacy in these islands, is different from the house rat in that it normally breeds at some distance from man. Human plague is said to have correlated with plague in house rats since it was only after the usual rat hosts had been killed by bubonic plague that the fleas sought the blood of man. The bite of the flea was the commonest mode of transmission of the bacillus into the human blood stream. This gave the bubonic type of plague. There were certain conditions, however, under which the plague bacillus entered the human body by way of the respiratory tract and engendered pulmonary or pneumonic plague. According to Shrewsbury[1] this form could not persist as an independent disease in the absence of the bubonic form.

Bubonic plague ('botch') was characterised by the appearance of buboes (swellings) of the lymph glands, particularly in the groin and armpits. In its final phase the infection took on the septicaemic form in which the bacillus passed directly into the blood. Pneumonic plague, characterised by spitting blood, was localised in the lungs and seems to have thrived under cool or cold conditions. It was directly transmissible from man to man through breathing, coughing, or sputum. This is not the place to offer hypotheses or explanations but it should be noted that 'the few medical historians who have taken cognizance of the detailed evidence of historic outbreaks of plague have tended to come to one conclusion: that bubonic plague was largely spread by *Pulex irritans* (the human flea) and not the fleas of rats!'[2]

Hirst[3] maintains that the distribution and density of rat populations governs the distribution and intensity of the human disease and that the rodent density is decisive because no serious outbreak of bubonic plague can take place in a locality supporting only a small or widely dispersed rat population. The house-rat population of Britain on the eve of the 'Great Pestilence' is not known, but it might be inferred from a study of the distribution and local densities of the human population about which a certain amount of information is available.

Rattus rattus found conditions congenial in the lowly dwelling houses crowded together in fourteenth-century towns (Fig 34). Stone-built houses of the well-to-do were not so congenial to them. Thus bubonic plague was primarily and principally a

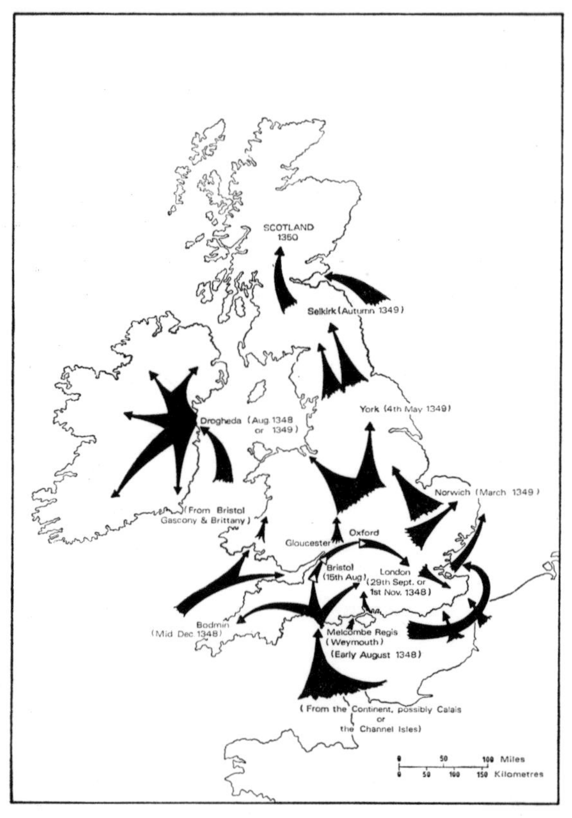

Fig 34 Progress of the Great Pestilence ('Black Death') in Britain in the fourteenth century (*based on texts of Creighton 1965, Ziegler 1968, and Salisbury 1970*)

disease of the poor. Where dwellings were scattered in the thinly populated countryside the rat population was too small for plague to become established and it struck only the occasional house here and there. In consequence the incidence of bubonic plague in fourteenth-century Britain was very uneven, and many places were spared its ravages. Shrewsbury's view is that 'in the comparatively densely populated region of East Anglia, and in the larger towns that were afflicted by it, "The Great Pestilence" may possibly have destroyed as much as one-third of the population; in the rest of England and Wales it is extremely doubtful if as much as one-twentieth of the population was destroyed by it. These are not random assertions; they are inherent in the aetiology of bubonic plague'. This might explain why the crisis of 1349 seemed to have had less of an effect on the military and political history of the period than on social and economic history.

Bubonic plague entered Britain at the little haven of Melcombe Regis—the modern Weymouth—in Dorset, early in August 1348. It came to Europe either *via* Calais or *via* the Channel Islands[4] and spread thence with a speed relative to the transport conditions of the day. Fleas were transported about the human person or in his belongings while the timid *Rattus rattus* was conveyed within bulky merchandise. Creighton says the disease spread from Melcombe Regis through Dorset, Devon, and Somerset, and reached Bristol by 15 August 1348 (Fig. 35). It was in Gloucester, Oxford, and London either 'at Michaelmas' (29 September) or 'at All Saints' (1 November).[5] It was active in London during the winter months of that year, an unusual occurrence because in temperate latitudes the rat fleas hibernate and the plague becomes quiescent or is extinguished. The disease also moved southwestwards through Devon into Cornwall. 'The deanery of Kenn to the south and south-west of Exeter is believed to have been the worst affected in the whole of England: eight-six incumbents perished from a deanery with only seventeen parish churches.'[6]

Norwich and the Eastern Counties were smitten in the spring of 1349 and suffered severely during the summer. York was attacked about the first week of May and suffered until the end of July. The spring and summer of 1349 were seasons of great mortality all over England except perhaps in the Southern Counties where the outbreaks began.

The disease apparently moved into Monmouthshire before it had run its course through Gloucestershire and Worcestershire. It seems to have travelled north along the Welsh border and

entered North Wales near Holywell. Thence the plague sprea
quickly in the summer of 1349 to such places as Ruthin an
Llangollen. The Black Death probably reached Carmarthen b
way of the sea.[7]

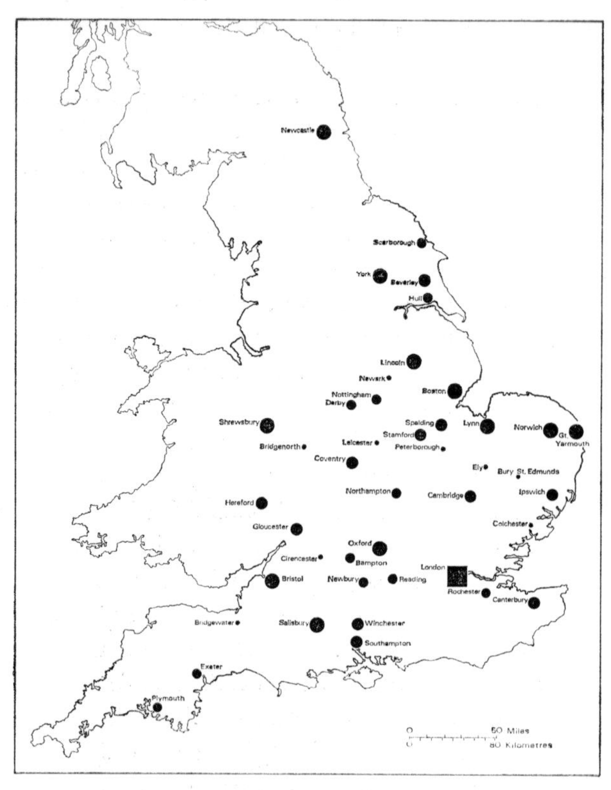

Fig 35 Ranking of the provincial towns of England, 1334 (*based on data in Hoskins 1960*)

Of the appearance of plague in Scotland in 1349, John o
Fordun wrote:

> By God's will this evil led to a strange and unwonted kind of death
> in so much that the flesh of the sick was somehow puffed out an
> swollen, and they dragged out their earthly life for barely two days

The 'pestilence' came to Scotland from Cumberland and Durham

Preſent Remedies

againſt the plague.

Shewing ſundrye preſeruatiues for the ſame, by wholſome Fumes, drinkes, vomits and other inward Receits; as alſo the perfect cure (by Implaiſture) of any that are therewith infected.

Now neceſſary to be obſerued of euery Houſholder, to auoide the infection, lately begun in ſome places of this Citie.

Written by a learned Phyſition, for the health of his Countrey.

Printed for Thomas Pauyer, and are to be ſold at his ſhop at the entrance into the Exchange.

1 6 0 3

Plate 5 Present Remedies against the plague. Title-page of a book published in London in 1603

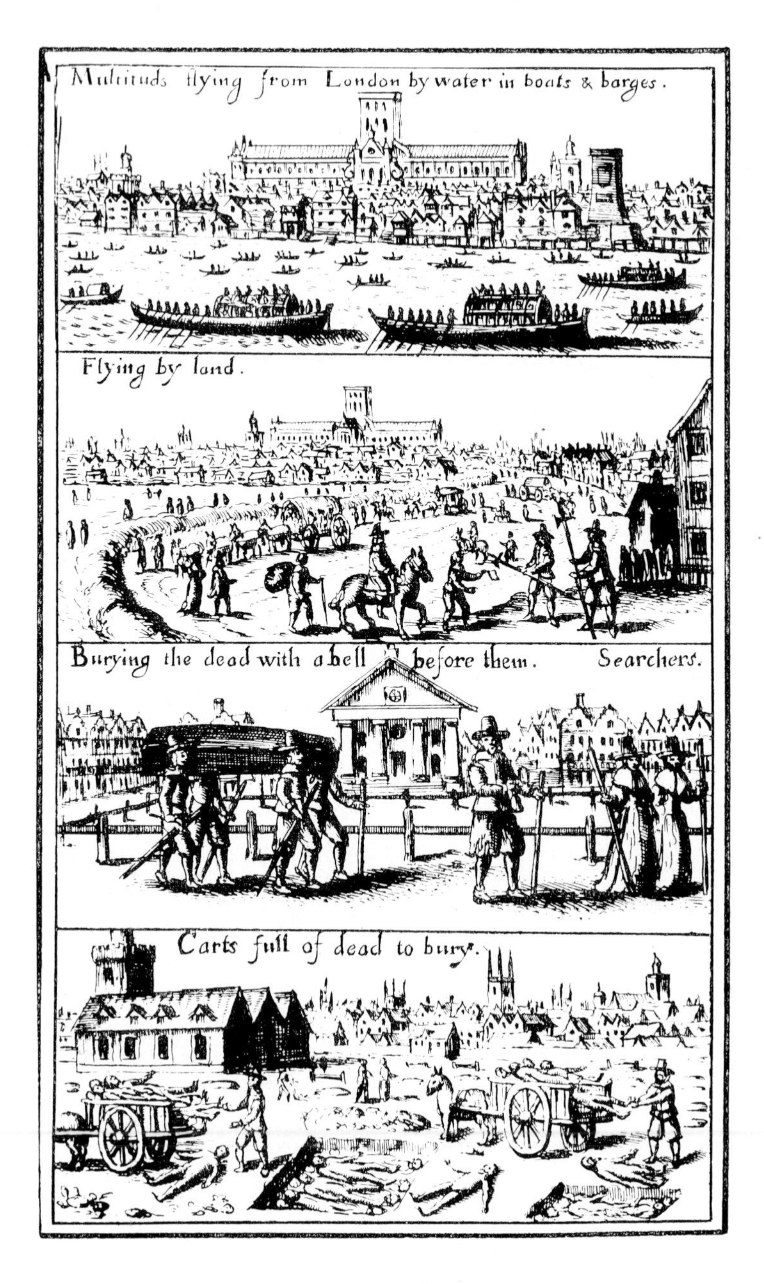

Plate 6 Scenes during the Great Plague of 1665

For a time its progress was held up at the Scottish border and 'the foul death of the English' gave the Scots malicious pleasure. But from a reckless foray into England the Scots brought the plague into their own country, where it was estimated that it destroyed one-third of the population. 'Men shrank from it so much that, through fear of contagion, some, fleeing as from the face of leprosy, or from an adder, durst not go and see their parents in the throes of death.'[8] That Scotland suffered severely is confirmed by Andrew of Wyntoun:[9]

> In Scotland, the fyrst Pestilens
> Begough, off sa gret wyolens,
> That it was sayd, off lywand men
> The thyrd part it dystroyid then
> Efftyr that in till Scotland
> A yhere or more it was wedand
> Before that tyme was nevyr sene
> A pestilens in our land sa kene:
> Bathe men and barnys and women
> It sparryed noucht for to kille them.

The disease made only slight progress in Scotland during the winter of 1349, but the following spring advanced with renewed vigour to encompass virtually the whole of the country.

'The Great Pestilence' in Britain extended over a period of three years: southern England in the latter half of 1348, the whole of England and Wales and the south of Scotland in 1349, and Scotland in 1350.

While plague spread to some remote hamlets its incidence and effect was nevertheless uneven. East Kent, for instance, was only slightly affected while West Kent experienced heavy mortality. In Dorset one hundred benefices were vacated in seven months. Deaths ran highest in sea and river ports and coastal districts where presumably rats obtained food most easily; they were lowest in pastoral or sparsely populated hilly areas. Stretches of marshland and fen acted as effective barriers against the disease.

Some authorities are of the opinion that the population of Britain was halved, others that it was reduced by 30 per cent; Shrewsbury carves the figure down to one-tenth at most. Studies by Rees[10] and others, of manorial documents, indicate that in some parts of England at least scarcely one-tenth of the inhabitants were left alive: 'Animals wandered about without an owner, goods lay open on all sides and the harvest remained ungathered.' In the period 1348–77 the population of England fell from about 3·8 million to 2·5 million.[11] London's population in 1377 was about 35,000. For York it was 11,000, Bristol 9,500, Coventry 7,000, Plymouth 7,000, and Norwich 6,000 (Fig 36).

The medical profession was virtually helpless in both the prevention or cure of plague. Later centuries produced a veritable flood of preventive orders, plague tracts, and defensive measures

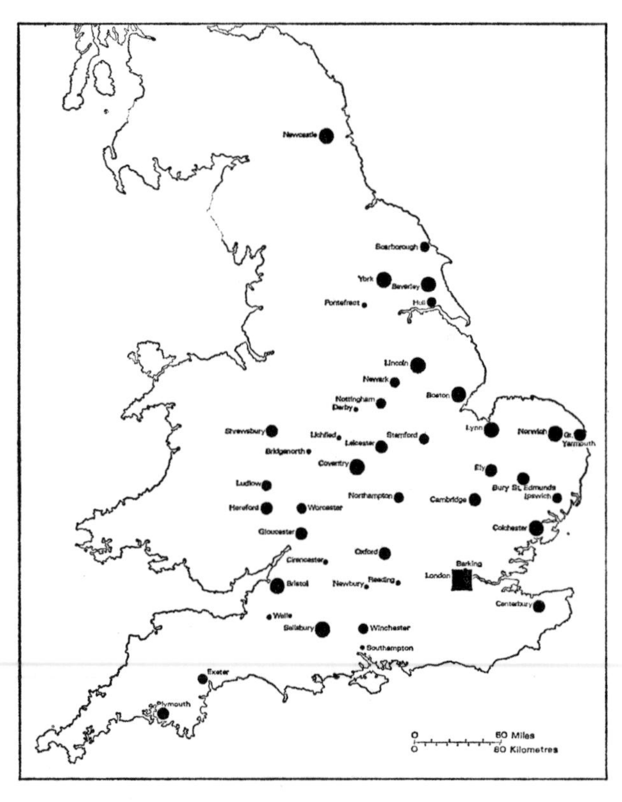

Fig 36 Ranking of the provincial towns of England, 1377 (*based on data in Hoskins 1960*)

(plate p 99) but the fourteenth century regrettably produced little practical activity on the part of contemporary British physicians. There is evidence of some slight appreciation of quarantine and of sanitation problems. For instance, in 1348 Gloucester cut off all intercourse with Bristol. Not improbably other towns adopted a similar practice. Some public authorities such as London devoted special attention to the cleaning of the town's ditches and streets and the carting away of filth.[12]

In addition to the sometimes severe mortality there were important economic and social consequences of 'The Great Pestilence'. For a time art, education, trade, and industry were paralysed. Norwich, centre of the Flemish cloth-weaving industry and sixth city of the Kingdom, took a generation to recover something of its former prosperity after being struck by the plague. There was an easing of the pressure of population and a retreat from agriculturally marginal lands. Labour was at a premium, and instead of peasants seeking land, landlords were obliged to seek workers and tenants. In many parts scarcity of labour and high wages contributed to a changeover from arable to sheep farming and in some cases to the abandonment of villages. The decay of feudalism was accelerated and the villain (agricultural serf) became yeoman farmer or wage-earner. The new class of yeomen rented their own farms and bought land. Others remained as free labourers able to demand high wages. Parliament was obliged eventually to pass laws in an attempt to stabilise wages. It would seem that plague contributed to a change in the status of England from a society based on personal service to a money economy dictated by the State. Agitations for higher wages and higher standards of living were persistent. Bean[13] says that the peasantry and artisans of late fourteenth- and fifteenth-century England did, in fact, achieve higher standards of living following a considerable rise in wages.

Some appreciation, expressed in modern terms, of the magnitude of the experience in which the people of medieval Britain had been involved during the time of 'The Great Pestilence' may be obtained from Thompson's analogy between its after-effects and those of World War I.[14] In both instances he says, contemporaries complained of 'economic chaos, social unrest, high prices, profiteering, depravation of morals, lack of production, industrial indolence, frenetic gaiety, wild expenditure, luxury, debauchery, social and religious hysteria, greed, avarice, maladministration, decay of manners'.

For three centuries or more after 'The Great Pestilence', plague was not long absent from one part of Britain or another but taken as a whole, the country was never again affected to the same extent as in the years 1348–50. A so-called 'Second Pestilence' (*Pestis secunda*) occurred in 1361–2. It seems to have inflicted in particular the young (hence *Pestis puerorum*) and the well-to-do classes. Shrewsbury[15] is of the opinion that this was not bubonic plague but epidemic influenza. There is reference to pestilence in Scotland in 1362 with symptoms and mortality similar to those of the 1350 outbreak. A 'Third Pestilence' (*Pestis tertia*) affected England in 1368–9, and a 'Fourth Pestilence' (*Pestis quarta*) in 1375–6 affected the south of England with an outbreak four years later in northern England and Scotland. *Pestis quinta*, said to have been comparable to 'The Great Pestilence' in its severity, spread through most of England in 1390–1 and into Scotland in the following year.

The seeming ubiquity of bubonic plague in Britain in the fourteenth century must not obscure the fact that other diseases afflicted the population at that time. Most of them were untreatable, spread more readily than bubonic plague and were usually just as lethal. These included fevers, fluxes, running scabs, boils and botches, burning agues, pocks and pestilences. All of these are mentioned by Langland.[16] 'Burning ague' was almost certainly typhus fever (*see* p 116), a winter disease associated with dirt and destitution. It flourished on the medieval custom of wearing the same underclothing (and therefore the same fleas!) from Michaelmas to Lady Day. Pneumonia undoubtedly occurred in epidemic form in the winter months. In all probability whooping cough, smallpox, measles, diphtheria, the enteric fevers and influenza also occurred in widespread and deadly epidemics.

Talbot[17] draws attention to one of the most curious epidemics to appear in the Middle Ages. The disease involved was called St Vitus's dance. 'Crowds of people were suddenly smitten with an urge to dance and though they professed to be suffering agonies whilst doing so, continued their compulsive movements for long periods at a time, even until they fell down and died.' He quotes one of the earliest and one of the most vivid accounts given by Giraldus Cambrensis of events at St Almedh's Church in Breconshire, South Wales.

> The saint's day was celebrated again in the same place where it had been celebrated for many years. The day was August 1st. Many people came here from distant parts, their bodies weakened by various diseases, hoping to be healed through the merits of the holy maiden . . . Men and women could be seen in the church and churchyard, singing and dancing. Suddenly they would fall down quite motionless, as if in a trance, and then as suddenly leap up again like lunatics to perform tasks that were forbidden on feast days . . . One man appeared to have a ploughshare in his hands, another urged forward his oxen with a whip. They accompanied these tasks with songs, but the notes were all out of tune. You could see one imitating a cobbler, another a carpenter; one pretended to be carrying a yoke, whilst another moved his hands as if he were drawing out thread and winding it into a skein. One man would be walking up and down weaving a net with imaginary thread; another one would sit at an imaginary loom, throwing his shuttle to and fro and banging the treadle with jerky movements. Inside the church (which is more surprising), you could see the same people offering gifts at the altar, after which they appeared to rouse themselves from their trance and recover.

During the fourteenth century Britain suffered frequent famines at which time hundreds of people are thought to have starved. The year 1371 was described as the 'grete der yere'. Another bad year was 1383 when harvests were poor and the fruit crop was badly affected. Many deaths were attributed to starving people having fed on unripe or bad fruit.

Unless it assumed truly catastrophic dimensions and caused terror and panic in the whole population, as in the case of plague, pestilence, and famine, disease and death are rarely mentioned in extant records. There is no mention of the common, if not almos universal presence of unrelieved pain from infected wound suppurating for weeks, bad teeth and toothache, or gastric upset following the eating of rotten foods. Expectation of life at birth for males was about thirty years, and after surviving the first and critical year it was still only thirty-four years.

Periods of privation and malnutrition must have greatly impaired the resistance of the common folk to disease. However the economic and social *sequelae* of this period are in course of re-examination.

Notes to this chapter are on pp 250–1.

8

Tudor Times

The 'Great Pestilence' or Black Death was followed by a long series of recurrent outbreaks of plague. The disease was endemic in Britain in the late fourteenth century and fifteenth century. There were at least twelve outbreaks between the Black Death and the accession of the Tudors which affected the whole country and eight outbreaks which appear to have been confined to London. There is little record of the almost annual ravages of plague in Tudor times, although it is known from letters and a few surviving Elizabethan plague-bills that these were considerable and constituted the most formidable medical problem of the day. Indeed, the official celebration of Queen Elizabeth's accession had to be curtailed because of an epidemic in London which caused the death of over 30,000 people. At first the visitations were on a national scale but afterwards became more localised, being restricted to towns and, more particularly, to the larger towns. The incidence of plague became increasingly urban at a time when most of the population lived in the countryside. Alongside plague were severe outbreaks of typhus, measles, syphilis, and 'English sweat'. There were repeated outbreaks of these major infectious diseases which tended to obscure the presence of the common, everyday ailments of the people such as rheumatism, toothache, and septic infections of the skin, which were accepted as the normal accompaniment of living.

The sixteenth century was a period when life for the ordinary man was hard and comfortless. Birth rates and death rates were high. Infancy and childhood were marked by fearful mortality and early life was plagued by several killing and disabling diseases. In epidemic or famine years, death rates exceeded birth rates and sometimes population growth was at a standstill or even declined. At best the population increased slowly.

Eighty per cent of the population lived in rural communities and, as in previous centuries, was confined to the more fertile areas. Changes in the agricultural system from arable to sheep farming led to the enclosure of estates by landlords, particularly in the south-east of Britain. A single shepherd was often all that was required where previously there had been a number of ploughmen. The resulting pressure on the labour market combined with the financial instability of the Government of the day and debasement of the coinage led to widespread poverty. The equivalent of the wage rate in terms of consumables reached a particularly low level by the end of the sixteenth century.[1] Town life and industrial life had not developed sufficiently to absorb the new class of vagrant and rootless labourers which was created. It was this great social problem which resulted in the passing of the first Poor Law Act at the turn of the sixteenth century.

Towns were still relatively unimportant. The average-sized provincial town had perhaps 2,000 to 5,000 inhabitants. London about the year 1500 may have had about 60,000 people, York, Norwich, Bristol, and Edinburgh about 15,000 inhabitants each. Below these came a dozen or so towns such as Exeter and Salisbury each with between 5,000 and 10,000 people. At the turn of the century the population of England was just over 3¼ million, of Scotland about ¾ million, and of Wales, ¼ million. A total for Britain of 4¼ million in 1500 rose to between 6 or 7 million by 1600 (Fig 37). Expectation of life at birth was still little more than thirty years.

Methods of keeping livestock had made little progress up to Tudor times. It remained difficult to provide adequate winter food for the animals and it was still customary to slaughter and salt most of the non-breeding stock in the late autumn. Meat was in short supply during winter, although for a time the pigeon-loft provided a useful source of fresh meat for certain of the gentry along with the birds and beasts that were hunted and trapped.

Milk could not be kept for any length of time so a great deal of it was churned into butter which, though salted, was to a greater or lesser degree rancid by the time of eating.

Apart from London, York, and a few other centres, food for the main towns could be obtained without a great deal of difficulty from the immediate neighbourhood. There was no means of preserving perishable foods apart from drying, salting, or pickling. This was serious in the towns, where carcasses of meat or consignments of fish might be several days old before being disposed of. Indeed a contemporary theory on disease was that the smell from stale fish and stinking meat gave 'an odour of putrefaction' which led to outbreaks of plague.

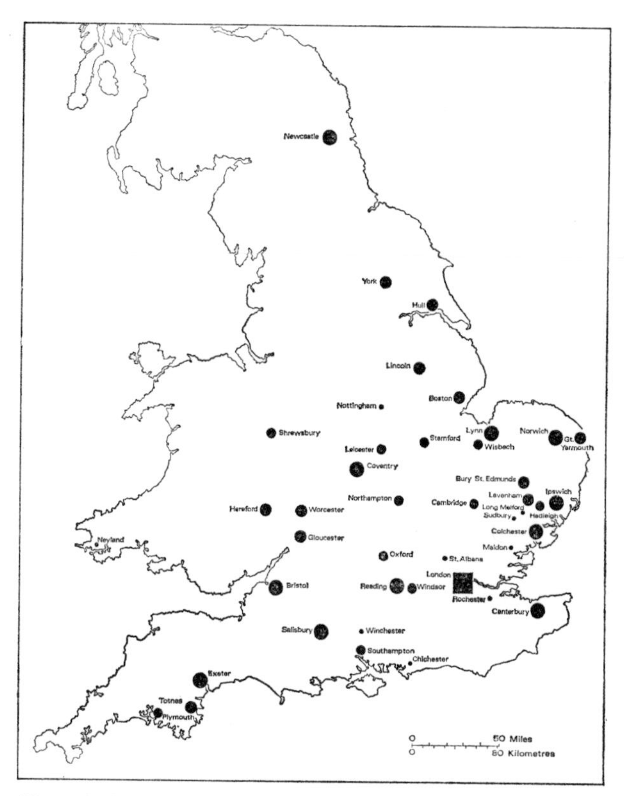

Fig 37 Ranking of the provincial towns of England, 1523–7 (*based on data in Hoskins 1960*)

Land enclosure in some areas meant the loss of grazing lands for peasants. They were thus unable to keep the occasional cow to provide them with 'white meat', as dairy produce was called in the sixteenth century. During the second half of the sixteenth century bread was the mainstay of the diet of the English village labourer, supplemented by peas and beans, soups, trapped game, and fish. Eggs were generally plentiful, sometimes supplemented by bacon or ham. It is said that the traditional English breakfast of bacon and eggs dates from this period. Oatmeal was an important item of diet in Scotland. Perlin (1551–2) after observing that Scotland was plentiful in provisions 'which are as cheap as in any part of the world', wrote that the poor people 'put their dough between two irons to make it into bread and then made it into what was esteemed good food in that country and tolerably cheap'.

On the whole diets were deficient in vitamins A and D and this is thought by Drummond[2] to account for the then common occurrence of stones in the bladder and urinary tract. Vitamin A deficiency might also have accounted for such eye diseases as night blindness (xeropthalmia).

There were several years of dearth during the second half of the sixteenth century when malnutrition, if not actual starvation, must have been widespread.[3] Periods of food shortage in the latter part of the sixteenth century were frequently followed by outbreaks of plague and typhus. Indeed, Saltmarsh[4] says that 'more than any single catastrophe this continual sapping of the human resources of England would account for the gradual but continuous decay of her national prosperity.'

The sixteenth century saw the beginnings of statistical records in the form of the London Bills of Mortality and the parish register. The Bills began probably in 1532 and were published thereafter at irregular intervals, usually only when plague was epidemic (plate p 66). Between 1563 and 1592 the regular series began. The Bills were compiled by Parish Clerks[5] from lists furnished every Tuesday by 'Searchers' whose duty it was to inspect the dead and register the cause of death. 'The Searchers hereupon (who are ancient Matrons sworn to their office) repair to place where the dead Corps lies, and by view of the same, and by other enquiries, they examine by what Disease or Casualty the Corps died.'[6]

The reliability or otherwise of the lists drawn up by the Searchers (who were seldom able to diagnose the cause of death except in obvious and familiar conditions as trauma or plague), is not relevant here. The Bills, nevertheless, provide some indication of the diseases prevalent at that time. Among these, in addition to plague, may be cited 'sweating sickness', syphilis, and 'gaol fever' (typhus).

According to Creighton, a confirmed believer in the localist–miasmatic theory of disease causation and opponent of the newly-formulated and rudimentary germ theory, 'sweating sickness' (English Sweat) suddenly appeared in Britain at the end of the fifteenth century. It was in Creighton's view a 'new' disease with a high mortality. Death occurred sometimes within six hours of onset, with sweating, from which the disease took its name, as a prominent symptom. Seemingly the disease remained localised in various parts of England and after five epidemics, in 1485, 1508, 1517, 1528, and 1551 respectively, vanished as mysteriously as it had come. The disease was known a few days after the landing of Henry VII at Milford Haven on 7 August 1485, and certainly before the Battle of Bosworth on 22 August. It broke out in London after Henry's arrival in that city and was severe at the end of September and during October, causing great mortality. Among the victims were two mayors and four aldermen.

Nothing was heard again of the sweating sickness until 1507–8 when a second outbreak of the disease occurred. This was less fatal than the visitation of 1485. A third epidemic in 1517 was more severe. Several towns, including Oxford and Cambridge, had a 50 per cent death rate. The disease recurred for the fourth time in 1528 and with great severity. It first showed itself in London at the end of May and spread quickly over the whole of England. Mortality was particularly heavy in the metropolis; the royal court was broken up and Henry VIII left the city, frequently changing his residence.

Sweating sickness spread over England and the rest of Britain, the Low Countries, Germany, and France on several occasions in the 1520s and there were further outbreaks in 1545, 1551, and 1558. The epidemic of 1551 was described with care by the physician John Caius.[7]

The sweating sickness, unlike plague, was not especially fatal to the poor but rather, as Caius affirmed, attacked the richer people and those who led a free life. 'They which had this sweat, sore with peril of death, were either men of wealth, ease or welfare, or of the poorer sort such as were idle persons, good ale drinkers and tavern haunters.'

Some attributed the sweating sickness, to the English climate with its damp atmosphere and fogs, others to the frightful lack of personal and domestic hygiene, still others to the intemperate habits of its victims. More recently it has been ascribed to food poisoning although a nutritional aetiology is also claimed as a predisposing cause. Neither of these was adequate, either separately or collectively, to produce the disease. The sweating sickness was probably a specific infective disease in much the same sense as were plague, typhus, scarlet-fever, or malaria. It was a clearly defined entity which ran a dramatic course. Contemporary descriptions of sweating sickness fail to reveal recognisable similarities to influenza as the latter disease is known at the present time, or to descriptions of an epidemic of influenza which occurred in Edinburgh in 1562. There were, in fact, influenza epidemics in Britain throughout the sixteenth century, particularly in 1510 and 1557–8, the last one comparable with the pandemic of 1918. The only modern disease resembling sweating sickness is that known in France as the Picardy Sweat, but there seems little doubt that the Picardy Sweat is not the same disease as influenza.

The disease arrived on the English scene after the Wars of the Roses and vanished as mysteriously as it came, never to appear again. In that time, however, it killed more than all the long years of the Wars had done.

There was still confusion in medieval times between leprosy, syphilis, and smallpox. True leprosy was declining at the end of the fourteenth century and by the fifteenth was no longer endemic, but the term 'leprosy', continued to be used later to describe a multitude of skin conditions. Creighton's numerous references to leprosy (of special note in this context is the Ordinance of Edward III which Creighton quotes and which, addressed to the Mayor and Sheriffs of London, states that lepers communicated their disease by 'carnal intercourse with women in stews and other secret places') were not leprosy but may well have been syphilis.

In the last years of the fifteenth century there was an extraordinary pandemic of syphilis ('French pox' or 'the Great Pox') in Europe. That it reached Britain there is little doubt[8] yet it makes hardly any appearance in records of that time. 'Wide and deep as the commotion must have been which caused it, it found hardly any more permanent expression than the private talk of the many of those days.'[9] The origin of the European pandemic remains a mystery. Some cite an American origin, believing that it was brought back from America by sailors who accompanied Columbus on his celebrated voyage. Others argue that syphilis had occurred in Europe from time immemorial and that up to the end of the fifteenth century and the beginning of the sixteenth century it had been a less acute and fatal illness and had been confused with leprosy and many other skin diseases.

There were indications of a somewhat unusual prevalence of *lues venerea* (*Morbus Gallicus* or *Morbus Neapolitanus*) in southern France in the autumn of 1494 and of a spread by contagion to Barcelona and Valencia in Spain. The expedition (1494–5) of the young French king, Charles VIII, which passed through southern France *en route* for Naples is said to have started the malady on Italian soil. There are also theories of a native Italian origin and also for a French origin for the disease. Hirsh has drawn parallels, on a minor scale, to cases of syphilis of severe type and communicable by unusual means, having been cultivated from quite commonplace beginnings among unsophisticated communities about the Adriatic and Baltic, and concludes that the 'mode of origin and the characteristics of these epidemics of syphilis appear to me to furnish the key to an understanding of the remarkable disease in the fifteenth century—an episode which entirely resembles them as regards its type and differs from them only as regards extent'.[10] At the present time there is doubt concerning a New World origin of venereal syphilis and yet it is not possible to prove for certain its presence in the Old World in pre-Columbian times.

Syphilis was thought to be a punishment sent by God and theologians maintained that general godlessness was the cause of the scourge. Doctors refused to have anything to do with the 'dirty' disease and handed over the sick to barbers, bath attendants, and charlatans. Syphilis is now known as a social disease, communicated by intimate contact, nearly always by sexual intercourse. The causative organism is the spirochaete *Treponema pallidum* (earlier *Spirochaeta pallida*) which has the important biological characteristic of requiring moisture for life and transmission. Continuous moisture is necessary for the transfer of the organism from one person to another. The socio-economic environment whereby the habits, customs, and attitudes of people permit them to have sexual relations with infected persons plays an important part in the transmission.

The epidemic of syphilis lasted for about 25–30 years in Britain, after which the disease reverted to the more endemic form of the present day. By the time William Clowes,[11] surgeon to St Bartholomew's Hospital, London, published his treatise on syphilis in 1579, the disease had lost the terrible severity of the original epidemic type.

Much points to Henry VIII having syphilis. The many miscarriages and still-born children of his Queens, his own chronic leg sores and a suspected nerve syphilis towards the end of his life suggest as much.[12] Syphilis was supposed to have been the reason for Edward VI's sickliness and early death, Queen Mary's childlessness and a contributory reason why Queen Elizabeth did not marry.

The early history of smallpox in Britain is obscure but it seems to have been a common disease in the sixteenth century. There was an epidemic of smallpox in 1561–2 and Queen Elizabeth herself suffered from a severe attack of the disease.

Medical teaching during the late Middle Ages continued entirely in favour of the Greek doctrine of Aristotle, Hippocrates and Galen.[13] This, a blend of observation, experience, and philosophy, was the humoral doctrine. The natural world was assumed to be made up of air, fire, water, and earth. Each of these elements had a characteristic quality; air was cold, fire was hot, water was moist, and earth was dry. A combination of any of these elements resulted in a blending of qualities or a complexion. Four complexions were recognised and with each was associated an appropriate humour.

Complexion	*Qualities*	*Humour*
Choleric	Hot and dry	Yellow or green bile
Melancholic	Cold and dry	Black bile
Phlegmatic	Cold and moist	Phlegm
Sanguine	Hot and moist	Blood

The complexion determined the individual's appearance and characteristics, but it could be affected by an excess of another humour. Thus a person of phlegmatic temperament would be rendered melancholic if anything occurred to cause an excess of black bile.

Ill-balanced humours were thought to predispose to disease in general, although it was admitted that some diseases such as plague, syphilis, ophthalmia, consumption (tuberculosis), and leprosy were exceptions. No theory had however been put forward to account for these exceptions and certainly no one knew what was transmitted. In 1546 Girolamo Fracastoro (Hieronymus Fracastorius, 1478–1553), who lived most of his life in Verona (Italy), published a small book called *De Contagione* which contains a theory by which it was thought that 'infection' was due to the passage of minute bodies, capable of self-multiplication, from the infector to the infected. Fracastoro was without doubt long before his time for his theory bears a superficial resemblance to modern doctrine. Indeed he might even be called the sixteenth-century pioneer of the germ theory. The term 'syphilis' originated with him for he also wrote a long medical poem entitled *Syphilus sive Morbus Gallicus* (Verona 1530).

There was much human misery in Britain in Tudor times. The

gaols were seriously overcrowded with unwashed wretches and the lice they brought with them sufficed to transmit typhus. There were at least three accounts of typhus ('gaol-fever tragedies'). They relate to the Cambridge Black Assizes in 1522, the Oxford Black Assizes in 1577, and the Exeter Black Assizes in 1586. In each case lawyers, county gentry and officials, jurors and others died. There is no mention of prisoners dying in the Cambridge episode but two or three prisoners died in chains a few days before the Assizes in Oxford. In Exeter there had been deaths in the gaol among Portuguese and English felons.

It was not known then that typhus (the causative micro-organism is a rickettsia—R. *prowazekii*) was conveyed from person to person by lice (*Pediculus humanus*). The lice were infected by feeding upon persons sick with the disease. Several explanations for typhus were offered; 'the savour of the prisoners' or 'the filth of the house' were invoked for the Cambridge Assizes; in Oxford it was the 'smell of the gaol'. In Exeter 'some did impute it to certain Portingals, then prisoners in the said gaol'. In the history of disease in Britain, typhus figures mainly in connection with the Civil War and as a disease of prisons, though, as noted later (Chapter 11), it was also common in the slum quarters of cities in the nineteenth century.

The winter diet of the peasant, made up of salty bacon, bread and peas and lacking in vitamin C, gave little protection against scurvy. In late winter and spring most of the poor country folk must have been at least in a pre-scorbutic condition. Drummond[1] says there is evidence to support the belief that pre-scorbutic conditions were common in sixteenth-century England. He cites the frequent mention in contemporary herbals of remedies for making 'loose teeth' firm and for 'purifying the blood in spring tyme'. The majority of the remedies were fresh herbs or extracts from them, fresh gooseberry leaves, raw purslane, elecampane leaves, raw gooseberries, decoctions of bramble leaves, leaves in wine, etc. Scurvy occurs when fresh fruit and vegetables are unobtainable or where people do not appreciate the need to include them in their diet. In the sixteenth century and later the most serious shortages of fresh foods occurred among sailors on

FUMIFUGIUM:

OR,

The Inconvenience of the AER,

AND

SMOAKE of LONDON

DISSIPATED.

TOGETHER

With ſome REMEDIES humbly propoſed

By J. E. Eſq;

To His Sacred MAJESTIE,

AND

To the PARLIAMENT now Aſſembled.

Publiſhed by His Majeſties Command.

Lucret. l. 5.

Carbonumque gravis vis, atque odor inſinuatur
Quam facile in cerebrum?——

LONDON:

Printed by W. GODBID, for GABRIEL BEDEL, and THOMAS COLLINS; and are to be ſold at their Shop at the Middle Temple Gate, neer Temple Bar. M.DC.LXI.

Re-printed for B. WHITE, at Horace's Head, in Fleet-ſtreet. MDCCLXXII.

Plate 7 Fumifugium: or The Inconvenience of the Aer and Smoake of London Dissipated. Title-page of a book by John Evelyn, first published in London in 1661

Rowlandson 1788-

Oh! this cursed Ministry! they'll ruin us, with their damn'd Taxes! why, Zounds! they're making Slaves of us all, & Starving us to Death!
by H Humphrey No 18 Old Bond Street
BRITISH SLAVERY.

long voyages and during military sieges. Under such conditions scurvy became a dreaded and much described menace. The true understanding of scurvy rests on the fact that man can survive several months without getting it when on a scorbutic diet since body stores of vitamin C are available. It is only after that period that the clinical manifestations of the disease occur, hence the increased incidence with the longer sea voyages.

Notes to this chapter are on pp 251–2.

9
Stuart Times

The chief epidemiological event of the seventeenth century was a series of outbreaks of plague and typhus. Historians have left the impression that no serious epidemic of plague attacked Britain between AD 1350 and 1603 and for the seventeenth century only the plague of 1665 has received special mention. Yet in almost any of the years 1348 to 1668 some community or other suffered visitations of these diseases and on several occasions considerable areas were devastated. Not surprisingly such ravages had an adverse effect on the economy of the country which in consequence suffered repeated setbacks.

In these early days plague, and epidemics generally, were assumed to be due to supernatural or astral causes, corruption of the air, the conjunction of Jupiter and Saturn or to popular superstitions such as ghosts, angels, coffins, hearses, flaming swords, and corpses in the air. Modern knowledge makes such theories appear ludicrous and the prophylactics futile (plate p 99) but an awareness and appreciation of such contemporary explanations or theories about disease is necessary for any real understanding of seventeenth-century attitudes towards the infection.

The London experience during the seventeenth century will suffice to illustrate the course of plague. The visitation of 1603 was believed to have been brought from Amsterdam though it had been reported in Cheshire, Derbyshire, and Lincolnshire the previous year. The epidemic attracted little or no attention before the end of April but thereafter, aided by crowds assembled for the coronation of James I of England (James VI of Scotland), and also hot weather, it acquired a firm grip on the city.

Public and private affairs were completely disorganised during that year. A proclamation postponed the coronation. Gentlemen were not permitted to attend court when it was in London and when business did not keep them in the city they were obliged to return home. No one was allowed to visit the royal palace or Westminster during the summer or autumn unless certified as having come from an uninfected area. Fairs were not permitted within fifty miles of the City, new building was prohibited and houses that had been recently built were pulled down. The London Companies were forbidden to hold public feasts in their Halls and it was suggested that a proportion of the money saved should be given towards the relief of the infected poor. All public meetings, feasts, and assemblies were postponed. Under such conditions of privation many people took to drink and riotous living, others to the churches. As in previous centuries physicians and surgeons deserted the city in its hour of greatest need; so did wealthy citizens, aldermen, and justices. The Court, accompanied by a disorderly company, migrated from one part of the country to another and infected the places to which it went. Among these were Southampton, Winchester, Woodstock, and Wilton. Plague also put an end to the export trade, particularly of broadcloth, and was equally destructive to domestic trade. With the approach of the colder and shorter days of autumn and winter the disease lost much of its virulence.

That the disease was associated with the flourishing rat population of London was not appreciated until several centuries later and long after plague had finally left the shores of Britain. The question posed at present is whether the 1603 outbreak was due to a recrudescence of a long-standing, smouldering disease, or to the introduction of a new or more virulent or infectious strain of the bacillus *Pasteurella pestis* in goods from overseas. Certainly the years preceding 1603 had been years of economic depression. There had been a succession of bad harvests from 1594–8 and again in 1600. The price of corn has risen 40 per cent. There was widespread poverty and plague is well-known as a poor man's disease. It flourished among the ill-fed, ill-clothed and poorly housed, among those living in rat-ridden tenements and insanitary alleys. Rarely was a plague victim one of wealth or mark.

The London epidemic lasted from March 1603 to January 1604

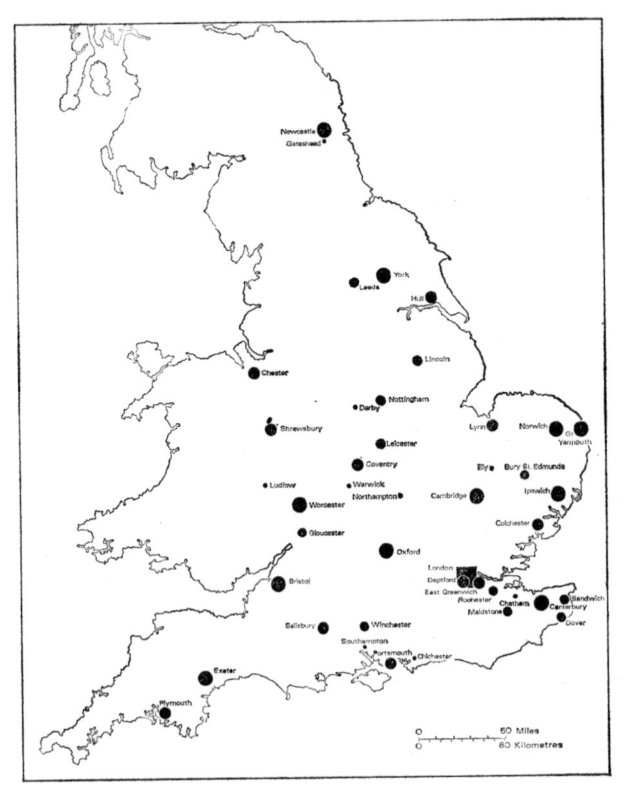

Fig 38 Ranking of the provincial towns of England, 1662 (*based on data in Hoskins 1960*)

and was responsible for the death of practically one-eighth of its quarter million inhabitants (Fig 38). Plague also raged fiercely in Bristol, Bath, Chester, Shrewsbury, Manchester, York, Enfield, Cranbourne, and Hassington (Northants) and parish registers record melancholy tales of households being entirely wiped out.

London was not free of plague for the next few years. All the time the disease was most deadly in the hot weather; in several years its activities were confined entirely to the autumn. There was another disastrous visitation in 1625, with evidence of a pneumonic component. The death toll exceeded that of 1603 by several thousand (41,313 as against 33,347). Like that of 1603 the infection was possibly introduced in goods imported from Holland. All the horrors of that epidemic were experienced and witnessed again, only this time to a heightened degree and the dislocation of commerce was even more severe than in 1603. The author of *Lachrymae Londinensis* summed up the lasting impression of the 1625 epidemic upon the minds of Londoners when in 1626 he wrote: 'to this present Plague of Pestilence, all former plagues were but pettie ones . . . This, to future Ages and Histiographers must needs be Kalendred the *Great Plague*.'

One other London plague of importance between 1625 and 1665 was that of 1636. During that year 10,400 out of total of 23,359 deaths were due to the disease. This particular plague led to, among other things, a mass of official correspondence on the subject of rag-gathering for the making of paper. It was felt that the unwholesome trade in street-refuse for rags spread plague. It was probably as a preventive measure against plague that 'the grant for gathering of rags' was recalled on 15 April 1639.

The words from *Lachrymae Londinensis* proved to be tragically wrong. It was the Great Plague of 1665 which was destined to be the worst, and last, of a centuries-long series of outbreaks in Britain. The epidemic had been anticipated by such auguries as spotted fever (typhus), pleurisy, and pneumonia which had flourished since 1658 together with the fact that it was practically thirty years since there had been a plague epidemic. The Great Plague might have represented a revival of a smouldering endemic disease, although the visitation was sometimes credited to infected merchandise routes through Amsterdam. The goods in question were probably cotton, the Levant being at that time the chief source of cotton piece goods and cotton wool.

Crowds of people had flocked to London after the Restoration. Many had not acquired the immunity of those citizens who had survived previous and repeated outbreaks of plague. There were also many more living in conditions more congested and insanitary then ever before. This encouraged rats, rat fleas transmitted the bubonic plague and, once it had broken out, its rapid spread within the city was inevitable.

The plague began unostentatiously in London in the winter of 1664–5, but by early summer it was well established. There were 68,596 deaths from plague in 1665 (Fig 39), 2,000 in 1666, 35 in

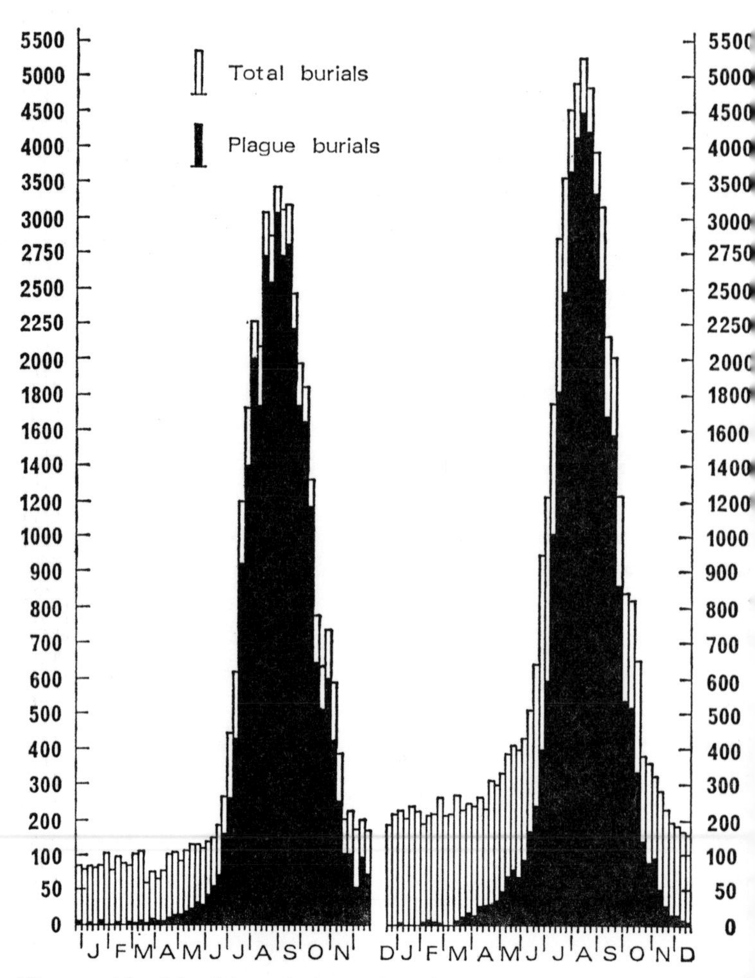

Fig. 39 Total burials and plague burials in London in 1603 and 1625 respectively (*after Shrewsbury 1970*)

1667, and 14 in 1668. Details need not be given since there are classic descriptions written by Samuel Pepys in his *Diary*,[1] and by Daniel Defoe in his historical novel *A Journal of the Plague Year*.[2] Some of the main features, summarised by Hirst[3] include the awful suffering of the poor; the paralysis of business; the constant tolling of the church bells and the frequent funerals in the early days, giving way, when the graveyards were full, to mass burial of corpses layer upon layer in the great plague-pits; the flight on foot, on horse, and on wagon of multitudes of panic-stricken citizens along the roads from London; all this forms one of the saddest chapters in the history of the English people' (plate p 100). Defoe's references to people falling dead in the streets and to death within a few hours of the appearance of 'tokens' suggest either the presence of septicaemic cases of plague or the lack of reliable clinical data. There were in fact remarkably few records of pneumonic symptoms or signs—which would be far more frequent in winter than in summer—during this or any of the seventeenth-century epidemics.

There were several accounts of plague in provincial towns in Britain (Fig 40). Among those affected were Southampton,

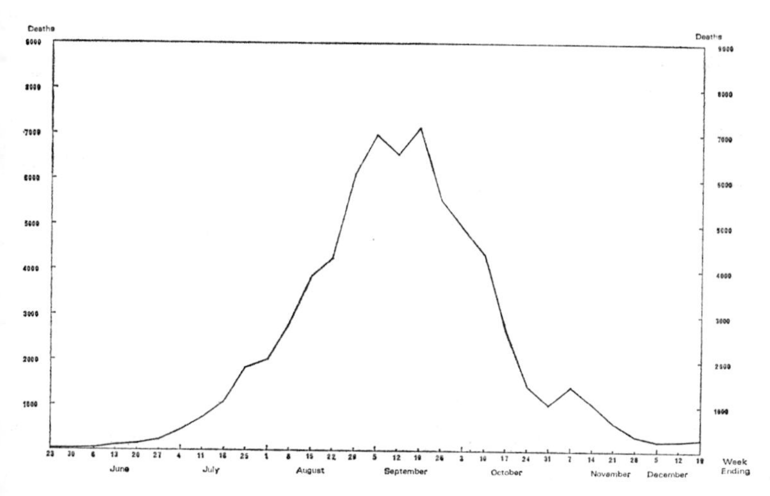

Fig 40 Deaths from plague in London, June–December 1665 (*based on text of Creighton 1965*)

Chatham, Cambridge, Yarmouth, Salisbury, Colchester, Norwich, Deal, Dover, Canterbury, Maidstone, Bristol, and Portsmouth. The South and Midlands of the country were hard hit, but apart from occasional pockets the North and West fared lightly. York, Newcastle, and Hull for instance escaped the disease and even within affected towns deaths were localised and restricted to a small number of families. Yet some towns and villages suffered severely from its devastating impact. One such was the small village of Eyam, during 1665–6.

> The plague was likewise at Eham, in the Peak of Derbyshire, being brought thither by means of a box sent from London to a taylor in that village, containing some materials of his trade. A servant who opened the aforesaid box, complaining that the goods were damp, was ordered to dry them by the fire, but in so doing it was seized with the plague and died.[4]

The first victim recorded in the Eyam parish register was George Vicars who died on the 6 September 1665; seemingly it was he who opened the box. Subsequent deaths followed a seasonal pattern consistent with other outbreaks of bubonic plague, a reduced mortality during the winter but rising to a climax the following summer. Several of the wealthier families from the west end of Eyam fled when the presence of the disease was discovered but over 300 persons remained. In June 1666, under the guidance of the Rev William Mompesson, the rector, and his nonconformist predecessor, the Rev Thomas Stanley, the villagers imposed their own quarantine on the village as a precaution against the plague spreading to neighbouring hamlets.

Oral tradition offers a guide to the measures taken to deal with the situation when it became serious in the spring of 1666. Men were still greatly influenced by the fifteenth-century idea that plague was not only infectious directly, as in the case of pneumonic plague, but that all who breathed the same air as the sick, or were exposed to their emanations, were carriers of the infection, domestic animals as well as man, and that inanimate objects of all kinds from an infected area could be sources of the disease. It was decided in June 1666, to immolate the village behind the famous *cordon sanitaire*, about half a mile in circuit, with its stones in which news and details of requirements were left and to which provisions were brought in accordance with an arrangement which was reached with the Earl of Devonshire, then at Chatsworth. The church was closed and services were held in the 'Cucklet Church', a natural formation in the Dell below the village. Burials no longer took place in the churchyard but in a variety of places around the village; an exception was made when Catherine Mompesson, the rector's young wife, died on 25 August. The Mompesson children, George and Elizabeth, aged three and four had been sent to relatives in Yorkshire in June and we have the rector's own testimony in his letter to his uncle of 20 November that he himself was unharmed: 'During this dreadful visitation I have not had the least symptom of disease, nor had I better health.' This same letter testifies to the erection of pest-houses in the village and to the use of chemicals on the sick, sometimes with efficacy: 'My man had the distemper and, upon the appearance of a tumour, I gave him some chemical antidotes which operated and after the rising broke, he was very well.' Letters were not sent directly from the village for fear of contamination: 'I have got these lines transcribed by a friend, being loth to affright you with a letter from my hands', Mompesson comments to his uncle.[5]

In the words of Creighton,[6] 'shut up in their narrow valley, the villagers perished helplessly like a stricken flock of sheep'. The village was decimated by the death of 259 of its inhabitants. In the light of present-day knowledge, flight from the village would have been preferable to the vain sacrifice behind the *cordon sanitaire*.

Scotland during the seventeenth century was an extremely poor country. The southern and wealthy part had been laid waste by wars with England in the middle of the sixteenth century, by internal political troubles associated with the period of the Reformation, by efforts made in 1650 on behalf of Prince Charles (later Charles II), by heavy fines subsequently imposed by Cromwell, and by plague. Plague continued in the south of the country in the early years of the seventeenth century after a long period of plague years during the second half of the sixteenth century. In 1606 'it raged so extremely in all the corners of the kingdom that neither burgh nor land in any part was free'.[7] After the storming of Newcastle by Scots Covenanters in October 1644 the plague appeared in Edinburgh, Kelso, Bo'ness, Perth Glasgow, St Andrews, Aberdeen, and other places. The disease was serious in Glasgow and made havoc from 1645 until the autumn of 1648.

The last case of plague in Scotland was in 1648; in the south-west and north-west of England it was about 1650; and in Wales probably 1636–8. The absolute last of its provincial prevalence in England was in Peterborough in the early months of 1667.

In the mid-twentieth century, at a time when there is a threat of mass human extermination by very rapid means, happenings during both the Black Death of the fourteenth century and the Great Plague of the seventeenth century shed interesting light on human behaviour under conditions of what, at the time, seemed to be universal catastrophe. A fundamental reaction throughout was flight.[8] Kings and their households, physicians, surgeons, merchants, lawyers, the clergy, professors, students, and the rich generally, were all involved in mass migration from the towns, leaving the ordinary folk to shift for themselves. Strenuous efforts were made to segregate those who were forced to remain in the towns by quarantining houses, closing off entire streets and the erection of gallows to warn against the violation of regulations.[9] Many people of all classes gave themselves up to carousing and ribaldry and even Samuel Pepys and his wife indulged in a 'great store of dancings'. London and Oxford experienced much 'lewd and dissolute behaviour'. There was also a wave of violence and crime. Some people were driven to a complete abandonment of morality, others resorted to religious extravagances. That there were profound disturbances of men's minds by ubiquitous and chronic grief and by the immediacy of death is unquestionable.

The main plague epidemics of the seventeenth century in London (including the liberties and out-parishes) and their associated mortalities were:

Year	*Plague deaths*
1603	33,347
1625	41,313
1636	10,400
1665	68,596

The popular nursery rhyme,

> Ring-a-ring o' roses,
> A pocket full of posies
> A-tishoo! A-tishoo!
> We all fall down.

takes its origin from the Great Plague.[10] A rosy rash it is alleged, was a symptom of the plague. Posies of herbs were carried as protection, sneezing was a final fatal symptom, and 'all fall down' was exactly what happened.

The Paracelsist physician Dr G. Thomson,[11] in his *Loimotamia* tract on the London Plague of 1665, describes how he cured himself of plague by applying to his stomach a large dried toad sewn up in linen cloth 'where after it had remained some hours, became so tumified, distended (as it were blown up) to that bignesse, that it was an object of wonder to those who believed it'. Toads were thought to exert a special influence on plague and could draw out the poison.

Examination of the Bills of Mortality which gave the probable reasons for the death of citizens of London in the seventeenth century shows that, in addition to plague, 'consumption and cough', 'ague and fever', 'cold and cough', 'quinsy and sore throat', 'flux and smallpox' figured prominently.

'Consumption' was the heading for various wasting diseases under which pulmonary tuberculosis (phthisis) appeared in the Bills of Mortality. It was, and had been, a common disease, accounting for 15 to 20 per cent of all deaths in London at the time, but it had not been epidemic in the sense of it being more prevalent at one time than at another. Tuberculosis is now known to be due to the invasion of the tissues of man or animals (especially cattle) by *Mycobacterium tuberculosis*,[12] but in early records tuberculosis was confused with other diseases. Even so, environmental conditions necessary for the active spread of the disease were undoubtedly present in areas of persistent overcrowding in close-packed dwellings. A low standard of living is known now to be another predisposing association. The susceptibility of the poorer people must have been greatly increased by malnutrition. Death rates from pulmonary tuberculosis in Britain are thought to have reached a peak about 1800.[13] For this reason more detailed consideration of the disease will be given in Chapter 11.

'Epidemic agues', 'hot agues', 'new agues', and 'quartan agues' were widely prevalent in different years of the seventeenth and earlier centuries, but they were not related to the endemic fevers of malarial districts such as the Isle of Sheppey, the Fens, and the Somerset Levels. Ague really referred to any acute fever and most commonly to 'continued fever' such as typhus (putrid malignant fever) or enteric (typhoid, or 'slow nervous fever').

Typhus was a constant visitor to London in particular, and less frequently to smaller towns of the country. During the Civil War there were severe epidemics of this disease. Worthy of mention was the experience in Reading (Berks) in 1643. The Royal army in the town had been besieged for eleven days by Parliamentary forces under Essex. When the beleaguered garrison surrendered, Essex found Reading infected and 'a great mortality ensued among his men'. Dr Thomas Willis (1621–75), Royal physician, wrote:

> In both armies there began a disease to arise very epidemical; however they persisted in that work until the beseieged were forced to surrender, this disease grew so grevious that in a short time after, either side left off and from that time for many months fought not until the evening, but with the disease, as if there had not been leisure to turn aside to another kind of death . . .

Among the causes mentioned were 'putrid exhalations from stinking matters, dung, carcasses of dead horses and other carrion'. In particular there was the filth of 'unshifted apparel' and the associated body lice which transmitted the organism *Rickettsia prowazekii*. The following year (1644) Tiverton (Devon), which had been occupied by both the Royal and Parliamentary armies, suffered a serious epidemic of typhus from August to November.

Seemingly there were no epidemics of typhoid, paratyphoid, or dysentery as there were for typhus, but since the causal micro-organisms for the enteric fevers and typhus were then unknown the diseases were assumed to be related. There is no doubt but that with bad water supplies and defective sanitation, typhoid, paratyphoid, and dysentery were endemic diseases during the seventeenth and earlier centuries.

The seventeenth century witnessed the rise to prominence of smallpox and measles.

In early English medieval writings the two diseases—*variola* and *morbilli*—were inseparable companions but by the time of the London Bills of Mortality in the first half of the century they were distinguished as independent diseases.

There were several smallpox epidemics in London, eg 1628 and 1634, and, according to later experience, a high mortality in London in any one year meant a general epidemic elsewhere in the country in that or the following year. This appears to have been so for the period following the Restoration, but details are lacking. Epidemics are known to have occurred in Taunton in 1658, 1670, 1677, and 1684, in Norwich and Halifax in 1681, in Cambridge in 1674, and Bath in 1675. The Duke of Gloucester and the Princess of Orange, both children of Charles I, died of smallpox within a few months of each other in the year of the Restoration (1660). Queen Mary died from the same cause in 1694. Smallpox reached its peak in Britain in the eighteenth century, and is discussed at greater length in Chapter 10.

Measles ('mezils') was first recognised as an independent disease by the English physician Sydenham. It was particularly virulent and fatal in seventeenth-century London and struck in epidemic form in 1664, causing 311 deaths. The Bills of Mortality record 295 deaths from measles in 1670[14] and 795 deaths in 1674. Thereafter there was a long interval of low mortality until 1705–6. There is no mention of epidemic measles elsewhere in the country in the seventeenth century.

That summer diarrhoea in infants ('griping of the guts' o 'convulsions') was common in London in the latter half of th seventeenth century and particularly in the populous working class liberties and outparishes is well testified by the works o Harris.[15] He refers to epidemics as follows: 'From the middle o July to the middle of September these epidemic gripes of infant are so common (being the annual heat of the season doth entirel exhaust their strength) that more infants, affected with these, d die in one month than in any other three that are gentle.' Ther was a serious mortality in 1669 and in subsequent years to 1672, i 1675, 1676, 1678–81, and 1688–9. Seemingly these years had ho dry summers and autumns and the diarrhoea incidence rose durin August and fell again in October. The climate fluctuated then a now and one year's weather might well have differed considerabl from long-term characteristics. There is no reason to suppos that there were not occasional exceptionally hot dry summers, e 1666, 1667, 1676. Pepys notes 5 July 1666 as 'Extremely hot . . oranges ripening in the open at Hackney', and John Evely (1620–1706) refers to the dry year of 1681 as follows: 'June 12 There still continues such a drought as has hardly ever bee known in England.' In 1684 he wrote: 'July 2. An excessive ho and dry spring and such a drought continues as is not in m memory.' In 1685 there was another drought and on 14 June Evelyn wrote 'such a dearth for want of rain as never was in m memory'. The causal relationships, direct or indirect, betwee infant diarrhoea and hot dry summers are not fully understoo although the consensus of opinion is that outbreaks were due t some particular kinds of bacillus coli, possibly fly-borne an possibly associated with animal manure.

The pandemics of 1889–92 and 1918–19 are milestones in th history of influenza in Britain. The disease was present in th seventeenth century, though it was not known by that name unt the eighteenth century. There were occasional outbreaks in th first half of the century but Creighton[16] makes particular mentio of influenza epidemics in 1657–9, 1661–4, 1675, 1678–9, 1688, an 1693. Influenza was given such names as 'new disease', 'hot ague' 'new ague', 'new fever', 'new ague fever', 'new pestilence', 'ne istemper', and in Derbyshire 'the new delight'. There were two atarrhal epidemics or of influenza proper in the spring of 1658[17] nd 1659 respectively set within a two- to three-year period of pidemic agues. Their over-all effects were afterwards viewed as 'little plague' and popularly spoken of as a warning of the Great lague of 1665. The epidemic of the spring of 1658 arrived uddenly in April 'as if sent by some blast of the stars'[18] after long winter of intense frost. Willis's explanation of the epidemic vas related to the constant north wind which 'checked the natural ction of the blood in spring'. Sydenham, known for the accuracy f his observations, considered the epidemic between 1661 and 664 to be malaria or 'intermittent fever'. Molyneux's account of he influenza of 1693 in Dublin states[19]: 'It spread itself all over ngland in the same manner as it did here, particularly it seized hem at London and Oxford as universally and with the same ymptoms as it seized us in Dublin, but with this observable ifference that it appeared three or four weeks sooner in London, hat is about the beginning of October.' There seems no doubt lso that influenza was present in Scotland in the seventeenth entury. As early as 1173 there was a reference in the *Chronicle Melrose* to a bad kind of cough, unheard of before, which affected lmost everyone from far and wide, 'from which pest' many ied.

John Evelyn reminds us of some of the endemic ills of seven-eenth-century London in his *Fumifugium*[20] (plate p 117), dedicated o King Charles II (Founder of the Royal Society). Evelyn, con-erned with the pollution of London's atmosphere by 'Sea-Coale' rought in from Newcastle for use in domestic fires and by brewers, dyers, soap-boilers and lime-burners' and in 'glass-ouses, foundaries and sugar bakers', wrote:

> And what is all this, but that Hellish and Dismall Cloud of SEA-COALE? which is not onely perpetually imminent over her head; For as the Poet,
>
> *Conditus in tenebris caligine coelum;* but so universally mixed with the otherwise wholesome and excellent Aer, that her *Inhabitants* breathe nothing but an impure and thick Mist, accompanied with a fuliginous and filthy vapour, which renders them obnoxious to a thousand

inconveniences, corrupting the *lungs* and disordering the entire habi of their Bodies; so that *Catharrs*, *Phthisicks*, *Coughs* and *Consumptions* rage more in this one City, than in the whole Earth besides . . .

London, 'tis confess'd, is not the only City most obnoxious to th Pestilence; but it is yet never clear of this Smoake which is a Plagu so many other ways, and indeed intolerable; because it kills not a once, but always, since still to languish, is worse than even Deat itself. For is there under Heaven such *Coughing* and *Snuffing* to b heard, as in the *London* Churches and Assemblies of People, where th Barking and the Spitting is uncessant and most importunate . . .

England, Wales, and Scotland supported about 4 million peopl at the beginning of the seventeenth century and 6½ million at th end, distributed in close relationship to productive agricultura lands (Fig 41). The greater part of the population was country bred and confined largely to lowland England, especially sout of a line from the middle Severn Valley to the Wash. There were however, some areas of greater density of population associate with the growing textile, iron, and coal industries. Most of th people lived at subsistence level or hardly above it, even by th low standards of the time. Expectation of life was now about 4 years, though Evelyn has the following to say about London which in 1603, had a population of 250,000, of 320,000 in 1625 460,000 in 1665 (Graunt), and 530,000 in 1690 (King):

. . . there is a waste of near ten thousand people who are draw every year from the Country, to supply the room of those tha *London* destroys beyond what it raises. Indeed the supply that th Town furnishes towards keeping up its own Inhabitants appears s very small to the ablest Calculators and most rational Enquirer[2] unto this subject that he owns he was afraid to publish the resul But without the use of Calculations it is evident to every one wh looks on the yearly Bill of Mortality, that near half the children tha are born and bred in *London* die under two years of age. Some hav attributed this amazing destruction to luxury and the abuse o Spirituous liquors. These, no doubt, are powerful assistants, but th constant and unremitting Poison is communicated by the foul Air which, as the Town still grows larger, has made regular and stead advances in its fatal influence.

The seventeenth century was a period of economic readjustment, particularly in respect of corn production (arable) and wool production (pastoral). High prices for corn favoured by some good summers encouraged many farmers to change to corn. This

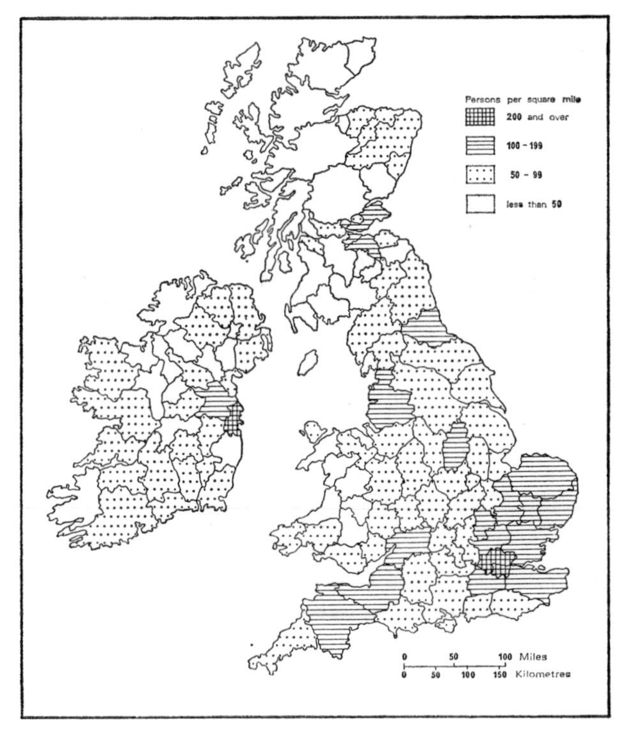

Fig 41 Distribution of population in the British Isles, 1670 (*adapted from Reader's Digest Atlas of the British Isles 1965*)

brought down the price of corn although with inevitable fluctuations after poor harvests. For the mass of the population times were hard and there was much unemployment and distressing poverty in the towns well into the middle of the century. Thorold Rogers writes: 'But I have never noticed in any earlier century such a continuity of dearth as from 1630 to 1637, from 1646 to 1651, from 1658 to 1661, from 1693 to 1699, in each case

inclusive.'[22] During such times the poor people suffered severel
from the high price of food. Indeed, according to Drummond,
the state of the lower classes in England during the latter part c
the century seems to have become progressively worse. The mai
food of the working class, according to contemporary account
was bread, beef, fish, home-brewed beer or ale, and cheese. Ther
was a tendency for people in the south and especially in the town
to give up dark rye and coarse meal breads for white bread (th
product of 'high milling' which removed the bran). In Lancashir
and the Pennines the peasants ate oaten cakes. Beans were repute
to constitute an important item of diet in Leicestershire and th
local population was twitted with the rude name 'bean belly'. I
Scotland lean years kept recurring with monotonous regularit
culminating from time to time in periods of acute starvation.
climax was reached in the dreadful famine of 1698–9.

Mention must be made of what was called a 'new disease' whic
came to Scotland in the middle of the seventeenth century. It wa
called *sibbens*[24] and was said to have been brought into the countr
by Cromwell's soldiers. It spread up the east coast at first an
reached Orkney and Shetland and then moved gradually south
wards to reach the Solway Firth. The similarity between sibben
and syphilis was noted from the beginning; subsequent researc
confirmed that sibbens was, in fact, syphilis.

Notes to this chapter are on pp 252–4.

10

Early Industrial Times

The eighteenth century in Britain saw the beginning of the change from the relatively simple agricultural economy of previous centuries, with local associations and peasant occupations, to a complicated industrial society with world-wide connections. It witnessed the introduction of steam (raised by coal) as a source of power for new mechanical inventions, the start of a new and intense concentration of large-scale industry on the coalfields and also of the labour force to man such industry. The eighteenth century was, in effect, the curtain-raiser to the Industrial Revolution and to the Steam Age of the nineteenth century. Above all it saw the trend towards a rapid growth in population which, continuing throughout the nineteenth century, is only now, in the late twentieth century, showing signs of slowing down.

At the turn of the eighteenth century Britain was still largely rural and agricultural (Fig 42). It exported some grain and certain raw materials but its only major industrial export was woollen cloth. A rather lengthy process of agricultural improvement had been taking place since the late sixteenth century, associated with such names as Jethro Tull, 'Turnip' Townsend, Thomas Coke, and Robert Bakewell. The Norfolk or four-course rotation in arable agriculture had been replaced by a three-year rotation and there had been a speeding up of land enclosure. Open fields had gradually disappeared and much moor and fen had been brought under cultivation. Enclosure caused local hardship, but the higher food production which followed improved agricultural techniques would not have come about under the previous open-field system.

Wheat, barley, clover, and beans were grown increasingly. Wheat yields in 1735 at about twenty bushels to the acre were twice those of the medieval period and exports continued until such time as the rise in population associated with developments in manufacturing industry absorbed all the food grown in the country. Root crops were grown to provide essential winter fodder for cattle which were no longer killed off in the autumn through lack of feeding stuffs.

The face of southern Britain had attained a mature agricultural–rural pattern by the eighteenth century and for the first fifty years fortune smiled on its people. There were occasional bad years, such as 1718, 1728, 1741, and 1757, but there were no long and lasting famines after the turn of the century. Improved

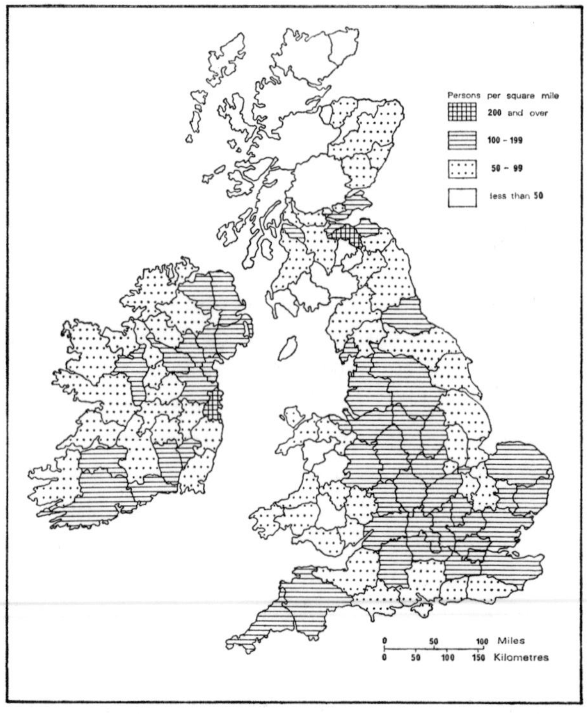

Fig 42 Distribution of population in the British Isles, 1750 (*adapted from Reader's Digest Atlas of the British Isles 1965*)

communications made it easier to transport food to the needy in the event of local crop failure. Food prices after 1730 were lower, relatively, than they had been previously and seemingly most people fed well and amply. Unless portrait painters and cartoonists such as Richardson and Gillray misrepresented their victims grossly, the consequences of gluttony were not apparently considered unaesthetic (plate p 118). There was also much heavy drinking among all classes of the community, particularly of cheap 'gin' and other forms of raw crude spirit. The overall impression is of an age of heavy eating and heavy drinking in both town and country (plate p 151). Artisans and labourers lived well and gout,[1] traditionally supposed to have been a disease of the rich and over-indulgent, was commonplace.

After about 1765 the poorer people knew hard times. The French wars (1756–63), a disastrous sequence of wet seasons and bad harvests during the period 1764–75, the interruption of trade (including foodstuffs and wines) at the time of the French Revolutionary wars, a financial crisis and a disastrous harvest in 1793 all contributed to a progressive worsening of conditions as the nineteenth century approached. The barest necessities of life became scarce and expensive, and discontent, which was widespread, occasionally culminated in lawlessness and riots. By the end of the eighteenth century a large proportion of the population was suffering depression, disaster, and death.

Population growth was relatively slow, increasing from $5\frac{1}{2}$–6 millions in 1700 to 6–$6\frac{1}{2}$ millions in 1750. Webster's[2] first census of Scotland in 1755 gave the population of that country as 1·3 million. Birth rates and death rates were high by modern standards and yet there were occasional years between 1720 and 1740 when the death rate exceeded the birth rate. After about 1750, however, there was an increase of population, variously ascribed either to a reduction in the death rate,[3] or a rise in the birth rate, or a shorter-run recovery or else 'compensatory fluctuation' from a rate of growth which previously had been abnormally low (caused by unusually high mortality earlier in the century (Fig 43)). Two-thirds of the people still worked on the land, distributed in close relationship with agricultural productivity. There were,

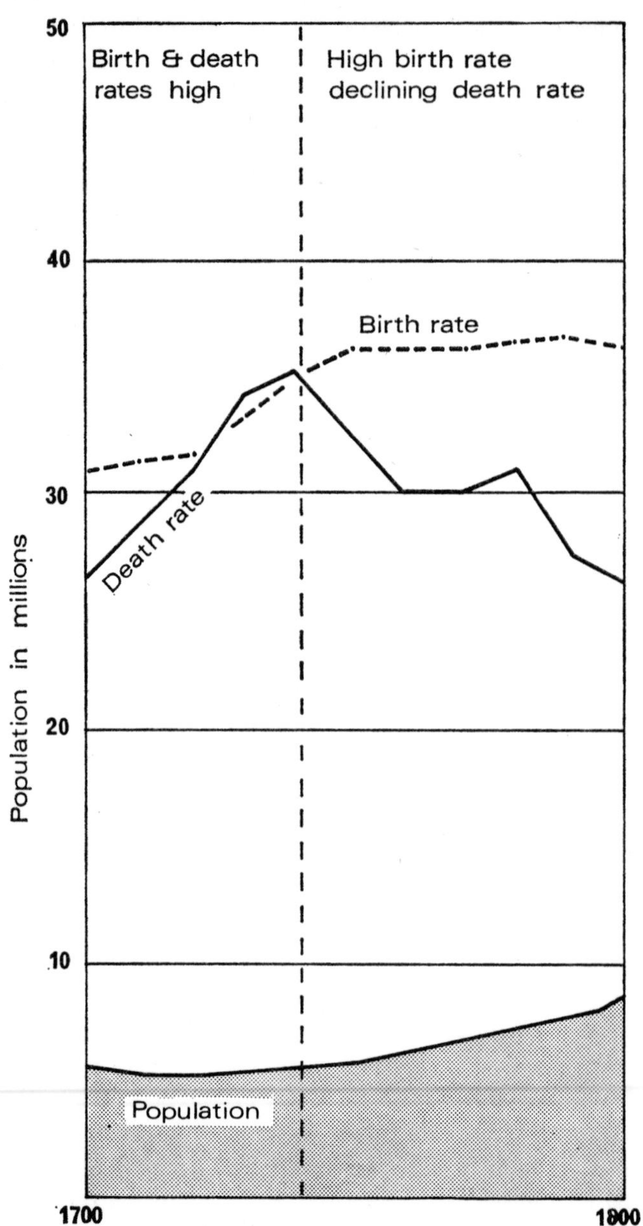

Fig. 43 Birth rates, death rates, and population totals in Britain, 1700–1800

however, areas of more dense population developing in parts of Yorkshire, north-east England and Lancashire. Dominating the populous areas of England were regional centres and market towns which included Plymouth, Exeter, Bristol, Birmingham, Coventry, Nottingham, Norwich, Hull, Leeds, Sheffield, York, Newcastle, Liverpool, and Manchester. London was by far the largest city and largest port. Bristol and Norwich were possibly second to London at this time. In Wales where the settlements were still small the main centres were the county and market towns of which the largest was Carmarthen with about 10,000 inhabitants. In Scotland, Edinburgh and Glasgow were growing rapidly, the Central Lowlands, especially the eastern side, having the highest overall density.

A quickening tempo of industrialisation occurred after 1750 but radical social and economic changes did not become particularly evident until early in the nineteenth century. Industrially, eighteenth-century Britain was not unimportant for she had woollen industries in East Anglia and the West Country, flax spinning and linen weaving in Renfrewshire and Lanarkshire, silk weaving in London, Coventry, Macclesfield, and Norwich, cotton in east Lancashire and in the Clyde Valley. Metallurgical industries were widespread and varied, and newer industries such as pottery, glass making, and paper making had developed considerably from the seventeenth century (Fig 44). Even so industry was still generally small scale and domestic in character with water the main source of power. Manufacturing industry was geared mainly to satisfying the demands of the home market and workers engaged in them worked under generally free and easy conditions, either in their own homes or in small workshops. It was not until water power was applied to bigger textile machines that the large factory became an element in the landscape. Even then the textile mills tended to be confined to more or less remote places. Similar considerations applied to Abraham Darby's iron-works at Coalbrookdale, Josiah Wedgwood's pottery at Burslem, Richard Arkwright's spinning mill at Nottingham, and the Lombe brothers' silk mill at Derby. The modern industrial landscape was only just beginning to take shape, the real blighting of

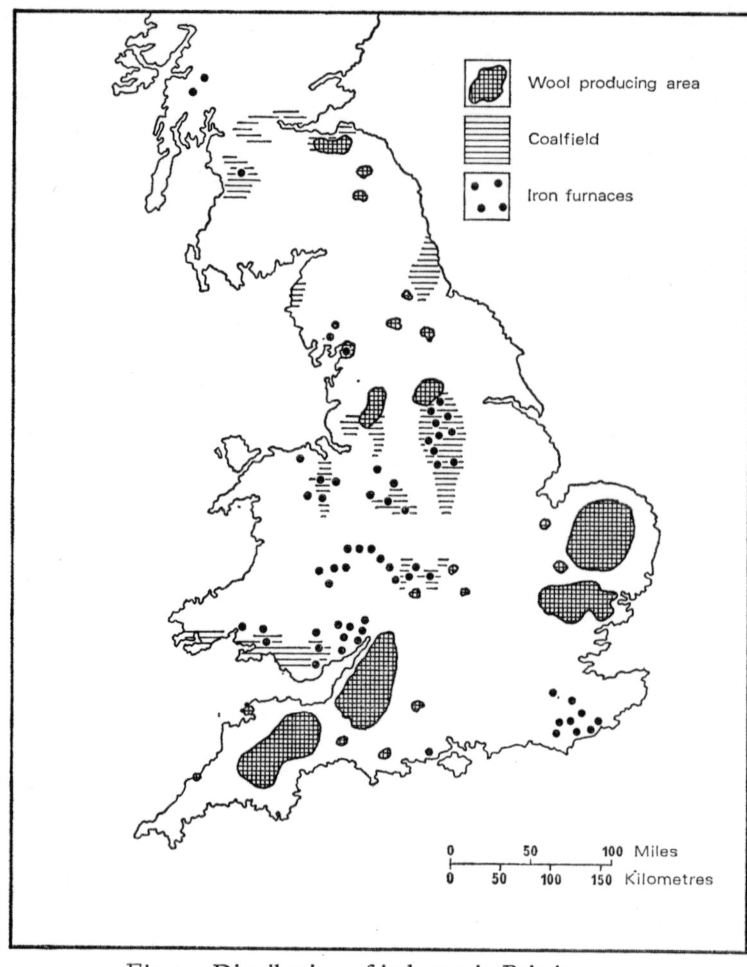

Fig 44 Distribution of industry in Britain, 1750

the environment with factories, industrial towns, atmospheric pollution, stream and river fouling in the coalfield areas came with the Steam Age of the nineteenth century.

So much for the industrial environment of eighteenth-century Britain. What of the diseases of the day, the maladjustments to environmental hazards? Plague, more or less endemic in Britain from the time of the Black Death in the fourteenth century to the Great Plague of the seventeenth century, had ceased. Why the

Great Plague should have proved to be the last of a centuries-long series of outbreaks has yet to be convincingly explained. The tremendous mortality, the 'sum of individual tragedies', of previous centuries was repeated in the eighteenth century, but this time the result of another deadly disease, smallpox.

Smallpox had been a scourge upon humanity for untold centuries. It had been described accurately by the great Persian physician Rhazes in the tenth century, although it had evidently been known in China and in India long before. Creighton, misusing the evidence relating to the early history of the disease in the sixteenth century, says, incorrectly, that smallpox in Britain came slowly into prominence and that it hardly attained a leading place until the reign of James I (1603–25). He omits from his narrative the epidemic of 1561–2, despite the well known fact that Queen Elizabeth herself suffered from such a severe attack of smallpox that it was thought by many at Court necessary to contemplate the appointment of a successor. Seventeenth-century epidemics in England, when young adults seem to have borne the brunt of the attacks, have been mentioned. There are also reports of an outbreak in Aberdeen in the summer of 1610: 'There was at this time a great visitation of the young children with the plague of the pox.'[4]

Smallpox was the most widespread and fatal disease throughout eighteenth-century Britain. Outstanding outbreaks occurred in 1722, 1723, and 1740–2 but the disease kept returning, time and time again, after relatively short intervals to the major industrial towns, after longer intervals to the market towns and after long intervals to the villages. Peak years in London were 1723, 1725, 1736, 1746, 1752, 1757, 1763, 1768, 1772, 1781, and 1796, each with over 3,000 deaths from smallpox. Edinburgh lost over 2,700 of its 40,000 inhabitants in the course of the two years 1740–2, more than half the deaths being of children under the age of five years. During the last quarter of the eighteenth century, nearly 19 per cent of all the deaths in Glasgow were due to smallpox.

The disease is a reaction between man and the smallpox virus or variola. It is an acute infectious disease which can be picked up anywhere but only if there are other cases or carriers of the infection in the vicinity. Cases of smallpox arise from contact direct or indirect, with a preceding case of the disease. The main mechanisms of infection is by the inhalation of infected droplets even then the virus is not ordinarily carried more than a few feet through the air. There are no natural insect or animal vectors, nor natural propagation of the virus outside the body. The virus ordinarily does not live long outside the body, yet seemingly it can be picked up from bed linen or clothing. The theory of the day was that the disease was the result of social and soil conditions.

Smallpox disfigured the faces of many of those whom it did not kill and caused much of the blindness of the eighteenth century. Thanks to keen observations and persistence in experimentation, Edward Jenner (1749–1823), a country doctor of Berkeley, Gloucestershire, gave to the medical world a method of combating the disease: vaccination. Early in his career Jenner had been impressed by the insistence of dairymaids suffering from sores and from mild reactions to cowpox that thereafter they would be safe from smallpox. Medical men believed this to be but an old country folk saying; but the idea intrigued Jenner. He collected examples of persons who had had cowpox and afterwards had escaped smallpox, or who, having had cowpox, did not react successfully to smallpox inoculation. Of this methodical work E. Ashworth Underwood[5] says:

> Even at this early stage he [Jenner] seems to have been obsessed by the feeling that cowpox *ought* to give complete and permanent immunity to smallpox. This is indeed strange, since every practitioner knew that smallpox did not always give complete and permanent protection against itself . . . Jenner set out to show that cowpox protected against smallpox, and also that cowpox could be transmitted from one human being to another just as smallpox could . . . that cowpox, naturally acquired, could be transmitted artificially from *person to person* so that there would result an increasing reservoir of persons who had been given the opportunity of becoming . . . immune . . . to smallpox . . . That was the cardinal factor in Jenner's doctrine and it was an idea which had probably not occurred seriously to anyone before; at least no one had attempted to put it into practice.

The first vaccination against smallpox was performed by Jenner in 1796. Exudate from a cowpox postule on the hand of dairymaid Sarah Nelmes was inserted into scratches on the arm of an eight-year-old boy, James Phipps. The vaccination proved to be effective protective therapy. Jenner was unable to repeat his successful discovery before 1798, in which year he published his famous *Inquiry into the cause and effects of the Variolae Vaccinae*. When it was realised that vaccination was a sure protection against smallpox the disease began to recede as an important cause of death and disfigurement in Britain (plates p 152 and 204).

'Fever', probably chiefly typhus,[6] was as important, if not more important than smallpox during the eighteenth century. It was a constant visitor to London in particular and less frequently to smaller towns. It was highly fatal in its epidemic form and was transmitted from person to person by the body louse (*Pediculus humanus corporis*). Lice took up the causative micro-organism *Rickettsia prowazekii* from the blood of people sick with the disease, and were themselves fatally infected in the process. The lice had about a week in which to transfer the infection to another subject before they died. Predisposing environmental conditions providing stimuli for the disease included the crowding together of poor, under-nourished, unwashed, and filthily-clad people. Typhus was especially common in the gaols and was often spread by contagion among court officers when prisoners were brought in for trial. 'Black assizes' comparable to those of Cambridge, Oxford, and Exeter in the sixteenth century (*see* p 116) took place in the Old Bailey in London in April 1750. The disease was normally unknown among the more affluent citizens but fifty or so people, including the Lord Mayor of London and several court officials, died following contact with dirty, neglected, and wretched prisoners.

It was ironical that at a time when most people seemingly enjoyed relative prosperity and general well-being an occasional poor harvest still gave rise to widespread distress. There were epidemics of typhus, called *synochus*, following bad harvests in 1718, 1728, and 1741. But in addition it was also a time of sloth, drunkenness, and thriftlessness. The particularly severe typhus epidemic of 1741–2 in London marked the climax of a number of years of severe fever mortality. The epidemic came after a long hard winter,[7] a dry spring, a hot summer, and a deficient harvest, although a complex of other environmental conditions also contributed to the spread of the disease. Not least was the gross overcrowding of dwellings, separated by narrow alleys and courts, especially in the City itself, sealed or deliberately blocked windows (resulting from the window-tax), cesspools beneath houses, stinking indoor privies, poor personal hygiene, drunkenness, low moral standards, and privation. It is small wonder that the London house in which Jonathan Swift then lodged had 'a thousand stinks in it'.

London mortality had never been so high as at this time. In 1741 typhus accounted for 7,500 deaths or about a quarter of the total deaths in the metropolis. Creighton described the disease as 'a curiously correct index' of the lowly condition of the working classes and the unwholesomeness of towns. It might have been appropriate but this description should be related to Creighton's own social, political, and medical beliefs. The 1741–2 epidemic, said to have reached Plymouth and Bristol from Ireland in the autumn of 1740, spread from there to Worcester and Exeter. Its subsequent diffusion throughout much of Britain might have been stimulated by the comings and goings associated with a General Election.

London was not alone of the big towns to be struck by typhus in the eighteenth century. Liverpool, Britain's second city (population 56,000 in 1790), housing 7,000 people in cellars and 9,000 in back-to-back houses, was equally troubled. So, too, were the poorer citizens of Edinburgh, Newcastle, Leeds, Hull, Carlisle, Lancaster, Manchester, Warrington, Chester and of the counties of Oxford, Gloucester, Worcester, Wiltshire, and Buckinghamshire. It is said that the death rate from typhus in Manchester in 1773 was just twice that of the surrounding countryside. This city had further outbreaks of typhus in 1794–5.

Scotland, like England, was sorely troubled by typhus. In 1741 for example following a year of famine, the disease inflicted itself on the other miseries of the people. In Edinburgh that year deaths from typhus amounted to nearly 20 per cent of all deaths. Then, during the 1745 Rebellion government troops returning from the Low Countries brought the fever with them and passed it on to the inhabitants in several parts of Scotland. Privation among the people during the latter half of the century was invariably associated with outbreaks of fever.

> Ye ugly creepin blastit wonner
> Detected, shunn'd by saint and sinner.
>
> *Ode to a Louse*, Robert Burns.

Cold weather, bad harvests, food scarcity, heavy drinking, filth, overcrowding, dirty clothes, and badly ventilated houses all contributed to the spread of the louse.

Enteric fever (typhoid, paratyphoid), one of the 'continued fevers' was, like typhus, familiar to everyone in the eighteenth century. It was a common, almost 'natural' cause of death, and in its endemic form was a constant menace and serious problem in public health. In the twentieth century enteric fever is generally spread by infection introduced from abroad[8] and may affect a wide area but in the eighteenth century it was pre-eminently a water-borne disease. The infection was spread by contaminated water (sewage) or by contaminated food but neither water nor food was the cause. The cause was a bacterium, *Salmonella typhi* (*Bacillus typhosus*) and the fever was due to the ingestion of the micro-organism in water or food. Not until the late nineteenth century when pure water supplies became available, sewers were substituted for privies, and more personal hygiene and cleanliness indulged in, did typhoid decline.

Up to the beginning of the nineteenth century it was virtually impossible to diagnose the nature of the many throat infections mentioned in medical writings. In particular it was difficult to disentangle scarlet fever from measles, erysipelas, or diphtheria. Fothergill[9] and Huxham[10] described epidemics of sore throat in London (1746–8) and Plymouth (1750–1) respectively, but Fothergill's description was of diphtheria and Huxham was speaking of a more or less concurrent outbreak of scarlet fever. In Scotland it may have been that cases of 'putrid sore throat' were in fact cases of diphtheria. The disease was said to prevail chiefly in damp situations in cold and rainy seasons, and more especially, near the sea; Stirling and Cupar in Fife were said to be particularly vulnerable. Both diphtheria and scarlet fever were serious infectious diseases in the eighteenth century, especially of childhood, and were a frequent cause of death. It is now known that diphtheria ('croup') is caused by *Corynebacterium diphtheriae*. There are three types, *intermedius*, *gravis*, and *mitis*, the two former are severe forms, the latter mild. Predisposing environmental conditions doubtless included overcrowding in people's work-places and in their homes.

Scarlet fever, known also as scarlatina, occurred in epidemic form on several occasions. The disease had been clearly differentiated by Sydenham in 1676: 'The skin is marked with small red spots, more frequent, more diffuse, and more red than in measles. These last two or three days. They then disappear leaving the skin covered with brawny squamulae as if powdered with meal.' In the context of scarlet fever rash and scarlet were synonyms. Modern knowledge of streptococcal sore throat tells that the disease was due to infection with haemolytic streptococci spread chiefly by droplet infection. For young children living in the close and crowded homes of the poor, scarlet fever was especially common.

Creighton writes of sore throat distempers (diphtheria and scarlatina) in Edinburgh in 1733, in Devon and Cornwall in 1734 in London late in 1739,[11] Sheffield 1745, London (Bromley, near Bow) in 1746–8, St Albans 1748, in Cornwall in 1748, Kidder-minster 1748, Plymouth 1751–3, London 1776–8, Birmingham 1778, Aberdeenshire 1791, and again in London in 1796–1805.

Confusion persisted, also, between scarlatina and measles (mezils), but there were undoubted epidemics of measles ('black measles') throughout the eighteenth century. The more severe epidemics of measles came to those communities which had lost their immunity through having been free of the disease for a long interval. The Bills of Mortality for London recorded 319 deaths from measles in 1705 and 361 in 1706, mainly among children. These fatalities were related to one continuous epidemic extending from October 1705 to April 1706. Further epidemics occurred; those of 1718–19, 1733, 1742, 1755, 1758, 1763, 1766, 1768, 1778, 1786, 1789, and 1792 were outstanding. The main victims were infants and young children. Creighton says 'measles caused by its direct fatality not more than a sixth part of the deaths by smallpox in Britain generally'.[12] In Scotland the disease appears to have been more common than scarlatina. The view of the College of Physicians of Edinburgh at the time was that the heavy drinking of spirits on the part of parents together with their general ill-health rendered offspring weak, feeble, and distempered and pre-disposed to infectious diseases. The real cause of this extremely infectious disease is a virus which is apparently spread through the secretions of the eyes and respiratory passages of the infected.

A further serious complaint among children, already present in the seventeenth century (p 132), was infantile diarrhoea. It is not known if the causal micro-organism then was the same as that which causes fatal diarrhoea in infants in the present century. What is known is that it caused enormous mortality among children in the summertime, July and August. Drummond's view was that 'the mortality among infants was in no small measure due to the heavy spirit drinking in the towns, particularly in the first half of the century. Wet nurses were all too often gin tipplers and the practice of giving spirits to quieten the infants was very general.'[13]

In the second half of the century, poverty, dearth, and high food prices added to life's misery. Thousands of children were abandoned, others were farmed out to foster mothers, only to succumb after a few months' 'care'. Under the insanitary conditions of the poorer parts of towns, or in such places as workhouses where large numbers of people were herded together, it was small wonder that infant mortality was high. One modern theory for the high mortality rate among infants in the eighteenth century implicates horse manure. It holds that infantile diarrhoea was fly-borne disease and associated with heaps of horse manure, an ideal breeding ground for flies. That horse manure was present in the streets and alleys there is no doubt, but also present were muckheaps in private premises, dead dogs and cats, the entrails and bones of cattle thrown on the streets, gutters choked witl refuse and pigs fouling the streets. Everywhere there was filtl and the menace to health was manifest.

Whooping cough (chin-cough) was not only a common an distressing malady of children but also a consistent contributor t the mortality of children in the eighteenth century. In his *Treati on the History, Nature and Treatment of Chin-Cough*, Robert Watt, i 1813, recorded that 'Next to the Smallpox formerly, and th Measles now, Chincough is the most fatal disease to whicl children are liable'.

Influenza, sudden in appearance, equally sudden in withdrawa and rarely lasting more than a few weeks at a time, was common i Britain in the eighteenth century. There were outbreaks in 1712 1727–9, 1733, 1737, 1743, 1762, 1767, 1775, 1782, and 1788. On epidemic which reached Britain in April 1782, though no mor outstanding in terms of character or mortality than previous ones may be mentioned since it was selected as the subject of two collec tive inquiries. It formed part of a pandemic which started in Asi and spread westwards through Siberia into Russia by Decembe 1781. From there it attacked Germany and Finland in Februar 1782, Denmark, Sweden, and England in the April, and the France, Italy, and most of the remainder of Europe. In Englan its first foothold was Newcastle but it spread rapidly southward reaching London, the Eastern Counties, Surrey, Portsmouth Oxford, and Chester by May, Yarmouth, Ipswich, Devon, Liver pool, and York by early June. Northwards it was in Edinburgl in May, and Glasgow by June, and affecting the whole of Scot land by July. Some 75 to 80 per cent of the adult populatio was affected by the epidemic, yet old people and children wer generally spared. 'People are variously affected with it, witl swelled faces, sore throats, dizzy heads, coughs, violent pains an feverishness; for remedy it prescribed a decoction of 2 oz lint seed, 2 do. of Liquorish-stick bruised and boiled over a slow fir in a pint water to half do., the strained and mixed with 4 o powdered sugar candy, also some lemon juice, brandy or rum take frequently a spoonfull thereoff etc.'

There was a progressive increase in the death rate from con

Plate 9 'Gin Lane' by Hogarth (1751)

A scene of London life and industry in which tradesmen, craftsmen and labourers over-indulged in gin-drinking. The statue of St George's, Bloomsbury, is in the background.

Plate 10 'The Cow Pock—or—the Wonderful Effects of the New Inoculation' by Gillray (1802), inspired by the publications of the Anti-Vaccine Society

sumption (pulmonary tuberculosis) during the eighteenth century. Immigration to the developing industrial towns was taking place on a scale without precedent and the accommodation provided to meet it was hopelessly inadequate. Lack of transport facilities and the need to live reasonably close to places of employment led to overbuilding of sites and overcrowding of houses. Domestic overcrowding offered ideal conditions for mass infection of susceptible people. Tuberculosis developed into a chronic,

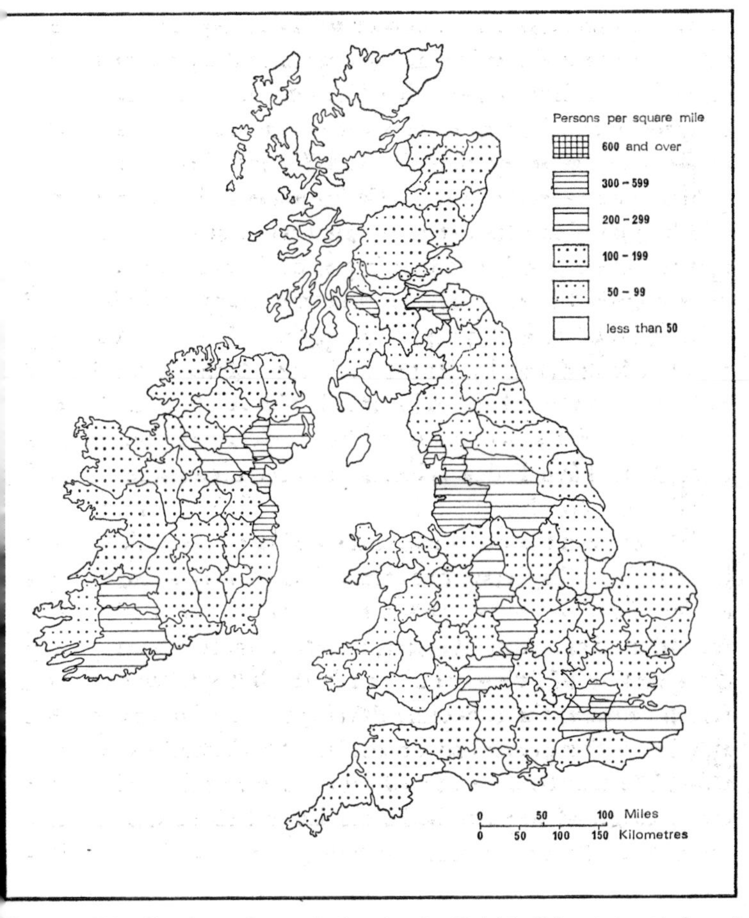

Fig 45 Distribution of population in the British Isles, 1801 (*adapted from Reader's Digest Atlas of the British Isles 1965*)

endemic disease. However it is as a scourge of nineteenth-centur Britain that consumption is remembered and so will receiv further consideration in Chapter 11.

By the time of the first census in 1801 the population of th British Isles had increased to 10½ millions (Fig 45). This increas has been variously ascribed either to a fall in the death rate or t an increase in the birth rate. Griffith[14] suggests that medica measures introduced during the eighteenth century had a sub stantial effect on the death rate. This view is not acceptable t McKeown and Brown[15] and they are probably right. These tw authors review the value of eighteenth-century surgery. In th case of midwifery they say that 'the introduction of institutiona confinement had an adverse effect on mortality', and that suc medicines as mercury, digitalis, and cinchona had no appreciabl influence on mortality trends. Of hospitals and dispensaries the said 'the chief indictment of hospital work at this period is no that it did no good, but that it positively did harm'. The onl disease upon which specific preventive therapy could have had substantial effect at this time was smallpox, but since vaccinatio was not introduced by Jenner until 1796 McKeown and Brow point out that 'it is hard to believe that inoculation can have bee responsible for a reduction in the incidence of smallpox larg enough to have had a substantial effect on national mortalit trends', and conclude that the rise of population was the result o an improvement in economic and social conditions. A substantia increase in the birth rate was considered unlikely to have occurre during the eighteenth century, except as a secondary result of reduction of mortality. It seems, therefore, that despite reports of deterioration of living conditions in the last quarter of the eigh teenth century, the economic developments of the period prob ably brought a general advance in the standard of living in thei wake. The effectiveness of medical therapy as a factor influencin the increase of population in the latter half of the eighteent century can be discounted.

Notes to this chapter are on pp 254–5.

11

Early Victorian Times

The age to which Queen Victoria gave her name was so long (1837–1901) that it cannot be thought of as one but as several ages. It was an Age of Steam. It was also an age of high birth rates, declining death rates and rapid growth of population, an age of social reform and substantial improvements particularly in real wages.

The rapid growth of population which had begun in the second half of the eighteenth century continued into the nineteenth century. At the time of the first census in 1801 the population of Britain was 10½ million; by 1831 it was 16·3 million. But not only did the population increase in numbers, its distribution was also changed in response to the developing cotton industry in Lancashire, the woollen industry of west Yorkshire, the knitwear industry of the east Midlands, and the mining and metallurgical activities of the coalfields of the Midlands. The Northumberland and Durham coalfields, the Yorkshire–Derbyshire–Nottinghamshire fields, those of the Black Country and along the Welsh Border had been exploited since the late 1500s, but were further stimulated during the nineteenth century. Instead of the populous zone extending in broad fashion across the southern half of England as in earlier centuries and related to a largely rural society there was now an axial belt from London to Liverpool related essentially to an industrial and urban society. London retained and even increased its national pre-eminence and with its suburbs had a population nearly one million by 1801. In contrast, eastern, southern, and south-western England (though showing substantial growth) had declined in relative importance. Bristol and its hinterland at the western end of the former zone of relatively dense population were eclipsed in importance by Liverpool the second port of the country and of the industrial hinterland of Lancashire. In fact Bristol was surpassed in population by

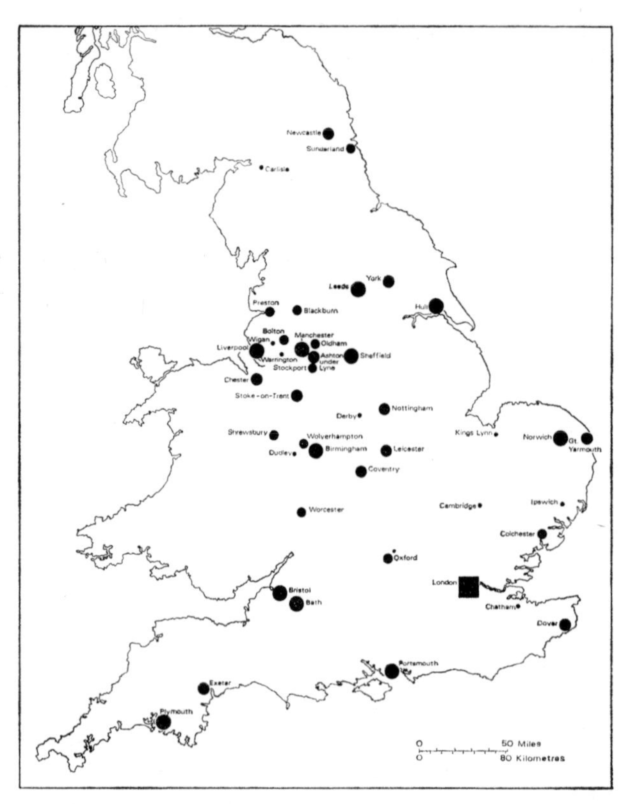

Fig 46 Ranking of the provincial towns of England, 1801 (*based on data in Hoskins 1960*)

Birmingham, Liverpool, and Manchester–Salford, although none as yet had populations over 100,000 (Fig 46).

Industrial developments had also taken place in South Wales and in the central lowlands of Scotland. In South Wales urbanisation expanded with developments in the metallurgical industries and in coal mining; in Scotland there was a gradual shift of emphasis from the east to the west in association with the very rapid growth of Glasgow and the surrounding district. Everywhere the new industrial towns created unprecedented social problems and hardships for the working classes.

The lack of employment opportunities for local population growth in the rural areas and a rising demand for labour in expanding industrial and mining areas resulted in a high volume of migration within the country. There was a general townward movement of the younger, active age-groups and a high rate of natural increase in the towns to which they went.

The class system in Britain which evolved from this movement of population was complex, but in broad terms resolved itself into two systems of authority, conflicting, complementary, or interdependent. The one was derived from the land and its traditional power, the other from developing industry. But the conflict was not one of equals; in the countryside there were the landlords as well as landless labourers; the towns had their industrialists and their factory operatives. The contrasts in experience of rich and poor were stark.

The struggle against Napoleon's France ended in 1815 with Britain in a thoroughly exhausted condition. Poverty, distress, and discontent were rife among the working classes. To the ill-effects of war and financial and trade disturbances were added the grave consequences of a sequence of indifferent harvests. The country came very close to real famine in 1812, but two bumper harvests in 1813 and 1814 brought the price of wheat down. Farmers protested that cheap corn would ruin home agriculture and the Corn Law of 1815 was passed to placate them. There was a considerable increase in unemployment in Britain following the closing of war industries after the French wars and this combined with three bad harvests in 1817–19 caused death, distress, and widespread rioting. The culmination was the breaking up by force of a public meeting in Manchester—the tragic Peterloo massacre—on 16 August 1819. The numbers of killed and wounded were disputed; 600 authenticated cases are known.

The proportion of population engaged in agriculture fell to about 30 per cent but productivity increased. Even so food-stuffs had to be imported in years of bad harvest. For the first twenty-five years of the century the conditions of the poorer people in both town and country remained bad. In Scotland many people were driven to desperation by the terrible struggle for existence after the repeal of the Salt Tax[1] and braved the hardships of long sea voyages to Australia and North America rather than endure the misery at home. There was particularly large-scale emigration from Britain to the New World and Australia between 1840 and 1850.

The large provincial industrial town was the unprecedented social phenomenon of early Victorian times. It usually grew from quite a small nucleus. Bradford, Yorkshire, a typical example, had little more than two streets at the beginning of the nineteenth century, each about a quarter of a mile long, converging on a bridge. According to the *Bradford Observer* for 9 June 1836 it was 'a town which previous to the introduction of machinery, was a miserable looking place, certainly not much more than five or six thousand souls, destitute of almost any public convenience or accommodation, its streets narrow and irregular, its buildings jumbled together without a design, or as if they had been dropped together by accident'. When it came much of its early population growth was absorbed within the existing nucleus of the town. Single or double rows of cottages were erected in the long gardens or crofts of existing houses, and reached by narrow passages frequently arched over. This irregular early development gave rise to 'the old slums'. The earliest industries established themselves on the edge of the original nucleus where all too many of them stayed. In the 1830s, developments on a larger scale began around the nucleus with the building of terraces of two-storey working-class houses—a typically English form of habitation. Back-to-back houses, were built in repeated rectangular blocks, perhaps 70 to 75 yards by about 35 yards, completely enclosing airless central courts reached only by tunnel entries. These soon merited the description 'the new slums'. It is possible to level criticism at such developments but it is important to appreciate the scale of the problem. Between 1841 and 1851 the population of Bradford increased from 66,718 to 103,786. And Bradford was no exception; the huddled courts in the original nucleus, surrounded by rows of back-to-back houses were repeated in practically every industrial town in the North of England.[2] In Scotland three- or four-storeyed tenements were more typical.

Conditions in these towns were worse than those in London, since few had any form of local government capable of ensuring even the most elementary standards of public hygiene. Many lacked the supply of such basic public amenities as water, sanitation, paving, and street cleansing, and harboured innumerable hazards to good health. Edwin Chadwick in his *Report on the Sanitary Condition of the Labouring Population of Great Britain* (1842) gave insight into the situation in these towns and provided disturbing evidence of the general living standards of the working people. In the *Report* one reads of ill-constructed houses, often mud-walled, with ill-ventilated rooms, neither boarded nor paved, and generally damp, of the retention of refuse inside the house in cesspools and privies. Near at hand were reeking dunghills and accumulations of filth; alongside habitations ran open drains oozing with sewage and animal and vegetable refuse, pigsties attach to dwellings, rubbish thrown alongside dwellings, and open slaughter-houses where the refuse and filth was allowed to accumulate for weeks without removal. In this way were the slums born.

Descriptions, in their own words, of the experiences of men, women, young people and children, who lived in and through the Industrial Revolution in Britain, are given in the following selection of quotations.

England's Manufacturing Population—
Housing Arrangements

One of the circumstances in which they are especially defective is that of drainage and water-closets. Whole ranges of these houses are either totally undrained or only very partially . . . The whole of the washings and filth from these consequently are thrown into the front or back street, which being often unpaved and cut up into deep ruts allows them to collect into stinking and stagnant pools, while fifty, or more even than the number, having only a single convenience common to them all, it is in a very short time completely choked up with excrementitious matter. No alternative is left to the inhabitants but adding this to the already defiled street, and this leading to a violation of all those decencies which shed a protection over family morals.[3]

Manchester

The greater portion of those districts inhabited by the labouring population . . . are untraversed by common sewers. The houses are ill soughed (drained), often ill ventilated, improvided with privies, and in consequence, the streets which are narrow, unpaved and worn into deep ruts, become the common receptacle of mud, refuse and disgusting ordure . . .

. . . surrounded on every side by some of the largest factories of the town, whose chimneys vomit forth dense clouds of smoke, which hang heavily over this insalubrious region.[4]

Liverpool

As in all large towns, the borough of Liverpool has its share of courts and alleys for the working population. Many of these, constructed within the last 10 or 15 years, are open, and afford comfortable dwellings, are well drained and clean, and are consequently healthy; others are of a very different class . . . The houses are generally built back to back, but this does not present a sufficient ventilation, as each has three openings, viz. a door, a window, and a chimney; some of the courts are closed, some open; the cleansing of the courts is left to the inhabitants themselves, and some of these will be found to be as clean as the most favoured parts of any town, whilst others, inhabited by the idle and dissolute, are as filthy.[5]

Birmingham

The courts of Birmingham are extremely numerous; they exist in every part of the town, and a very large portion of the poorer classes of the inhabitants reside in them. . . . The courts vary in the number of the houses which they contain. From four to twenty, and most of these houses are three storeys high and built as it is termed back to back. There is a wash-house, ash-pit, and a privy to the end, or on one side of the court, and not frequently one or more pigsties and heaps of manure. Generally speaking, the privies in the old courts are in a most filthy condition.

. . . a few of the circumstances in which Birmingham, perhaps differs from most of those large towns in which fever, constantly prevails, and in which its ravages are so formidable. These are—the elevated situation of the town—its excellent natural drainage, and its abundant supplies of water—the entire absence of cellars used as dwellings—the circumstances of almost every family having a separate house—and lastly, the amount of wages received by the working classes, which may be regarded as generally adequate to provide the necessities of life.[6]

Bradford

In some streets a piece of paving is laid half across the street, opposite one man's tenement, whilst his neighbour contents himself with a slight covering of soft engine ashes, through which, the native clay of the subsoil is seen protruding, with unequal surface, and pools of slop water and filth are visible all over the surface. The dungheaps are found in several parts of the streets, and open privies are seen in many directions.[7]

Sheffield

Sheffield is one of the dirtiest and most smokey towns I ever saw. . . . One cannot be long in the town without experiencing the necessary inhalation of soot, which accumulates in the lungs, and its baneful effects are experienced by all who are not accustomed to it. There are, however, numbers of persons in Sheffield who think the smoke healthy.[8]

Leeds

By far the most unhealthy localities of Leeds are close squares of houses, or yards as they are called, which have been erected for the accommodation of working people. Some of these, though situated in comparatively high ground, are airless from the enclosed structure and being wholly unprovided with any form of drainage, or convenience, or arrangements for cleansing, are one mass of damp and filth. . . . The ashes, garbage and filth of all kinds are thrown from the doors and windows of the houses upon the surface of the streets and courts. . . . The privies are few in proportion to the number of inhabitants. They are open to view both in front and rear, are invariably in a filthy condition, and often remain without the removal of any portion of the filth for six months . . .[9]

Nottingham

I believe that nowhere else shall we find so large a mass of inhabitants crowded into courts, alleys and lanes, as in Nottingham, and those too of the worst possible construction . . .[10]

Bath

. . . disease showing here and there a predilection for particular spots, are settling with full virulence in Avon Street and its off sets . . . Everything vile and offensive is congregated there . . . and to aggravate the mischief, the refuse is commonly thrown under the staircase; and water more scarce than any quarter of the town![11]

Greenock

In one part of Market Street is a dunghill—yet it is too large to be called a dunghill. I do not misstate its size when I say it contains a hundred cubic yards of impure filth, collected from all parts of town. It is never removed; it is the stock-in-trade of a person who deals in dung; he retails it cartfuls.[12]

Glasgow

It is my firm belief that penury, dirt, misery, drunkenness, disease and crime culminate in Glasgow to a pitch unparalleled in Great Britain.[13]

William Cobbet gives some indication of life in an English village. This is what he writes of Cricklade (Wiltshire):

> The labourers seem miserably poor. Their dwellings are little better than pigsties, and their looks indicate that their food is not nearly equal to that of a pig. Their wretched hovels are struck upon little bits of ground *on the roadside*, where the space has been wider than the road demanded. In many places they have not two rods (11 yards) to a hovel . . . Yesterday morning was a sharp frost; and this had set the poor creatures to digging up their little plots of potatoes. In my whole life I never saw human wretchedness equal to this; no, not even among the free negroes in America.[14]

Living conditions in early Victorian times were bad. But Dr Guy, a physician of King's College Hospital, London, giving evidence to the Health of Towns Commissioners, said that, bad as they were, the apartments of the poor were more wholesome than their place of work.[15]

Factories multiplied, first alongside streams and rivers but within the towns after steam power was adopted. It was in the towns of the industrial belt of the Scottish lowlands, Manchester–Yorkshire, the Midlands and South Wales that the British proletariat was born. Gaskell's description of workers employed in the great cotton mills reads as follows:

> Their complexion is sallow and pallid—with a peculiar pattern of feature, caused by the want of a proper quantity of adipose substance to cushion out the cheeks. Their stature low—the average height of four hundred men, measured at different times, and in different places, being five feet six inches. Their limbs slender and splaying badly and ungracefully. A very general bowing of the legs. Great numbers of girls and women walking lamely or awkwardly, with raised chests and spinal flexures. Nearly all have flat feet, accompanied by a down trend, differing very widely from the elasticity of action in the foot and ankle, attendant upon perfect formation. Hair straight and thin.[16]

Of the food habits of the early years of the nineteenth century the chief features were the increased consumption of potatoes and tea. Bread was still 'the staff of life'. It was white bread and all too often alum-whitened. In years of want much of it was baked with flour of very inferior quality, frequently mixed with potato flour, barley bran and other products which darkened the loaf. Alum was added to make it white. It was not only the bread that was adulterated. Copperas was added to beer, capsicum to mustard, wine was faked (much of it was made of spoiled cider), artificial tea was made from blackthorn leaves, and 'Radical Coffee' (so called because it was favoured by some of the disaffected) made from horse beans, rye, or wheat, partially carbonised and ground down.[17] The quality of milk too was unbelievably bad and supplies were heavily infected with tuberculosis. Much of the butter was rancid, meat was tainted and fish stank. Major regional differences in diet tended to disappear with the increasing dependence on purchased food, and the diet of the vast majority of town dwellers became more uniform.

The water supplies in many towns came from wells and many of these were polluted. Infected wells proved a serious menace to community health, witness the incident of the Broad Street pump in London (*see* p 177). Rivers too were often little better than 'elongated cesspools'.

Such, briefly, was the environment of early years of the Victorian age. Expectation of life at birth in 1841 was about 40 for a boy and 42 for a girl but with considerable regional variation and also differences between the social classes (Table 7).[18]

The main causes of death in early Victorian times were typhus, commonly called 'fever', smallpox, cholera, and tuberculosis. Lack of adequate protection against secondary infection also rendered some of the common fevers of childhood a serious threat to life, and infant mortality was high.

Typhus, still confused with the enteric group of fevers, was endemic and epidemic in Britain in the early nineteenth century. In previous centuries it had been associated with famines and undernourishment, now it was the constant accompaniment to life in the courts, closes, and wynds of the industrial towns. It was the poor man's disease ('that unerring index of destitution') the product of squalor, insanitation, overcrowding, and verminous conditions, a concomitant of working-class housing. It was persistent and devitalising. It smouldered but occasionally broke out with renewed virulence whenever a new crop of susceptible people came within its reach. There were outbreaks in 1817–19, 1826–7, 1831–2, 1837, and 1846–8.

The epidemic of 1817–19 visited almost every town and village of the United Kingdom. The fever was generally typhus in England but in Scotland it was to a large extent relapsing fever. The outbreak in 1826–7 had a distinctly more relapsing character

Table 7 Average age of deceased persons, by social class, in selected localities in England in early Victorian times

Localities	*Professional persons or gentry and their families*	*Tradesmen and their families*	*Labourers, artisans, servants, etc. and their families*
Unions in the County of Wilts.	50	48	33
Kendal Union	45	39	34
Derby	49	38	21
Strand Union	43	33	24
Truro	40	33	29
Kensington Union	44	29	26
Whitechapel Union	45	27	22
Leeds Borough	44	27	19
Bethnal Green	45	26	16
Bolton Union	34	23	18
Liverpool	35	22	15

in Scotland, although this was not altogether unobserved in London, Bristol, or elsewhere in England. Glasgow, Edinburgh, and Dundee had an epidemic of typhus in 1831–2 and a steady prevalence thereafter. The year 1837 witnessed yet another climax in Glasgow and Dundee, followed a year later in Edinburgh and Aberdeen. The corresponding epidemic in England, 1837–8, was almost wholly typhus.[19] It led to the death of 6,011 people in London in the space of eighteen months. Other large towns affected—mainly in the latter half of 1837—were Liverpool, Manchester–Salford, Birmingham, Bolton, Sunderland, Leeds,

Sheffield, Bradford, Stockport, Dudley, Abergavenny, Wolverhampton, Newcastle, Wigan, Chorley, Swansea, Halifax, Macclesfield, and Norwich.

Typhus deaths in the four largest towns of England were:

	1838	*1839*
Manchester–Salford	627	416
Liverpool	573	358
Leeds	245	150
Birmingham	123	141

Several of the outbreaks of typhus were associated with the Irish immigrations which followed potato famines in that country. The disease was endemic in Ireland and in times of stress, as accompanied the failure of a harvest and food shortage, it almost invariably produced an epidemic. That of 1846–7 was by far the worst outbreak and was one of the consequences of the ghastly Irish Potato Famine. The very wet and cool summers of 1846 and 1847, and the unbroken expanses of susceptible cultivated varieties of potatoes in the Irish fields were ideal for the small fungus *Phytophthora infestans*, which came in from America, the original home of the wild potato, on some infested tubers and spread with great rapidity. For two years the potato harvest on which the peasants depended for their food supply failed almost completely. Local workhouses or emigration to Britain were the only alternatives to death from starvation. In no time the disease, which was called 'famine fever' spread to the remainder of the British Isles, with dire consequences in the overcrowded cellar-dwellings of Liverpool and the slum tenements of Glasgow.

The years 1847–8 were particularly bad for typhus in Britain, and Lancashire and Cheshire, and Liverpool, Manchester, Birmingham, Dudley, Wolverhampton, London, Shrewsbury, Leeds, Hull, York, Sunderland, Edinburgh, and Glasgow were seriously affected. The experience in Glasgow, as described by Chalmers, will serve as an illustration:

> From 1816, indeed, until the early seventies of last century, the closes and wynds of the city were devastated by recurring epidemics

of infectious diseases of several kinds, and of considerable magnitude. Nor did these stand alone; they formed only the higher peaks of an elevated table-land of disease, which was capable of maintaining an annual death-rate, oscillating frequently between 30 and 40 per 1,000, and of rising in occasional years, under the influence of epidemic prevalances to 46, as in 1832, during the first cholera epidemic, and 56, as in 1846, when typhus fever alone caused a death-rate approaching 14 per 1,000 or only a little lower than the average rate for all causes at the present time.[20]

Glasgow was possibly the filthiest and unhealthiest of all the towns of Britain at this period. Immigration into the city had occurred on a scale without precedent and the accommodation provided to meet it was hopelessly inadequate. There was a general lack of transport facilities and this, coupled with the need to be living in reasonably close proximity to places of employment, led to overbuilding of sites and overcrowding in houses. Nor was there an effective system of refuse removal.

The pollution of water supplies and of gross surface impurity near houses was inevitable.

Pauperism, or destitution worse than pauperism, which demanded relief and failed to obtain it, was not only much greater in Scotland than in any other European countries similarly situated, but that it was greatly increasing, and that this increase together with the influx of rural and Irish pauperism, into our great towns, had brought them into a condition greatly more favourable than they had ever been before for the spread of epidemic disease and had accordingly raised their mortality far above the level of corresponding towns in England and on the Continent.[21]

The situation in other cities was little better. In Liverpool for instance there were, even as early as the decade 1787–96, an average of 3,000 or more typhus cases each year.[22]

Cholera presented a picture which was very different from that of typhus. This was a disease of rapid onset, dramatic course, highly lethal while it lasted, and exceptionally contagious. The pathogenic agent, *Vibrio cholerae*, on reaching the intestine causes diarrhoea, which can be fatal within two to six days through acute dehydration of the affected person. It is passed from man to man, either by stools which contaminate clothing, linen, or the hands allowing transmission through contact, or more usually throug the intake of water or food contaminated by the excrement o cholera patients.

Cholera spread throughout China, Burma, and Ceylon durin the eighteenth century propagated largely by pilgrims, militar personnel, and traders. In the nineteenth century there were som great epidemics occasionally developing into pandemics whic originated in an endemic source in Asia—essentially Lowe Bengal and Indonesia but with secondary sources in India- Pakistan, Burma, and southern China. Descriptions of thes Asian epidemics had been given by European settlers but choler did not spread westwards before the nineteenth century.

The disease reached global significance with the pandemic o 1817–23. It originated in lower Bengal but soon became th scourge of monsoon Asia, spreading eastwards to China an Japan as well as westwards through Mesopotamia to the shore o the Mediterranean and southwestwards across the Indian Ocea to the east coast of Africa. The infection spread both by sea an by old overland routes and left a terrible trail of death in China Ceylon, Burma, and Persia before subsiding, having failed t reach north-west Europe.

Britain became involved in the second and greatest pandemic This flared up in the valley of the Ganges in India in 1826 ad vanced rapidly through the Punjab and into Afghanistan an Persia to south-eastern Europe. Soon it spread to all parts of th western world (Fig 47).

When William IV opened Parliament on 21 June 1831 h announced: 'It is with deep concern that I have to announce t you the continued progress of a formidable disease in the easter parts of Europe . . .'

The advance of the epidemic westwards was followed b people in Britain with fear and trepidation. The epidemic reache Moscow by 1830 and survived and continued through a ver severe winter. It reached Berlin in August 1831 and Hambur two months later. As Hamburg was, at that time, a mere 36 hour away from England by steamship, the occurrence of choler

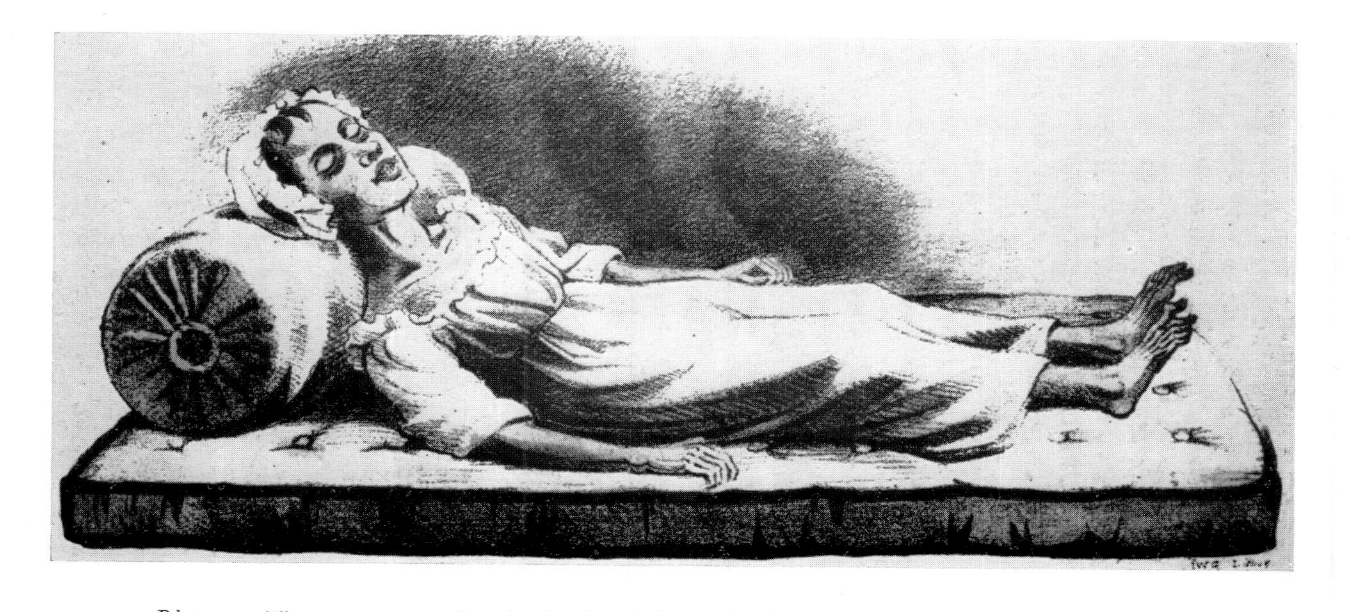

Plate 11 The appearance after death of a victim to the Indian cholera who died in Sunderland

Plate 12 Tombstone in St Michael's Kirkyard, Dumfries, to 420 victims of cholera, 1832

there provoked considerable public concern in Britain. Ministers were urged in the House of Commons to impose stricter quarantine and to set up a Board of Health. On 18 October 1831, the Privy Council issued a set of *Instructions and Regulations regarding cholera*, prepared for them by the Board of Health.

It was in Sunderland, on 19 October 1831, that cholera made its first appearance in Britain and the first case was diagnosed on 4 November (plate p 169). From Sunderland it moved on to

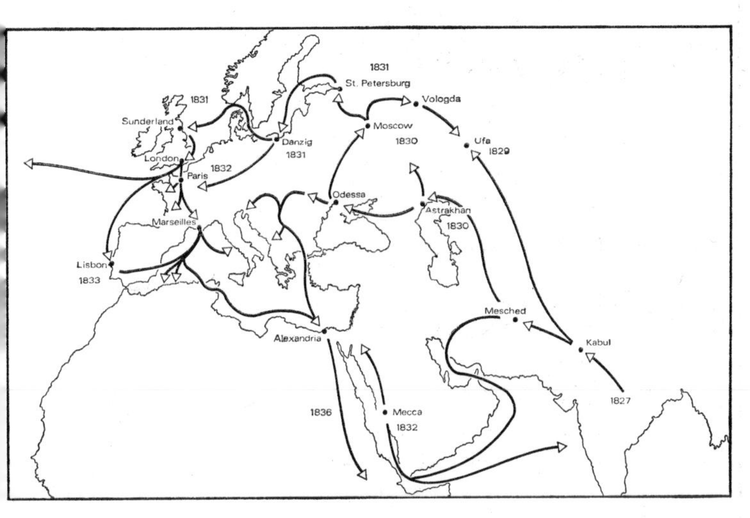

Fig 47 Progress of the cholera epidemic from India to Britain, 1827–31 (*adapted from Rodenwaldt 1952*)

Tynemouth, Newcastle, and Gateshead and to villages within a few miles of them on both banks of the Tyne. The disease then spread northwards through Northumberland into Scotland, arriving at Haddington, East Lothian in December, Hawick, Roxburgh, and Edinburgh in January 1832, Glasgow early in February, and thence northwards into the Scottish Highlands. The heaviest death roll in Scotland occurred in Glasgow where there were 3,166 deaths (plate p 170). Another wave of the disease moved southwards to engulf in turn Leeds (700 deaths, Fig. 48), York (200 deaths), Liverpool (1,500 deaths), and Manchester–Salford (900 deaths). London was reached by February but was free of cholera by the autumn, the official figure of deaths for the year being just under 5,300. The Midlands suffered a heavy attack in June with over 2,000 deaths by the end of November. The disease was rampant amid the shacks and hovels of the new industrial districts. There was a brief outbreak in Sheffield from

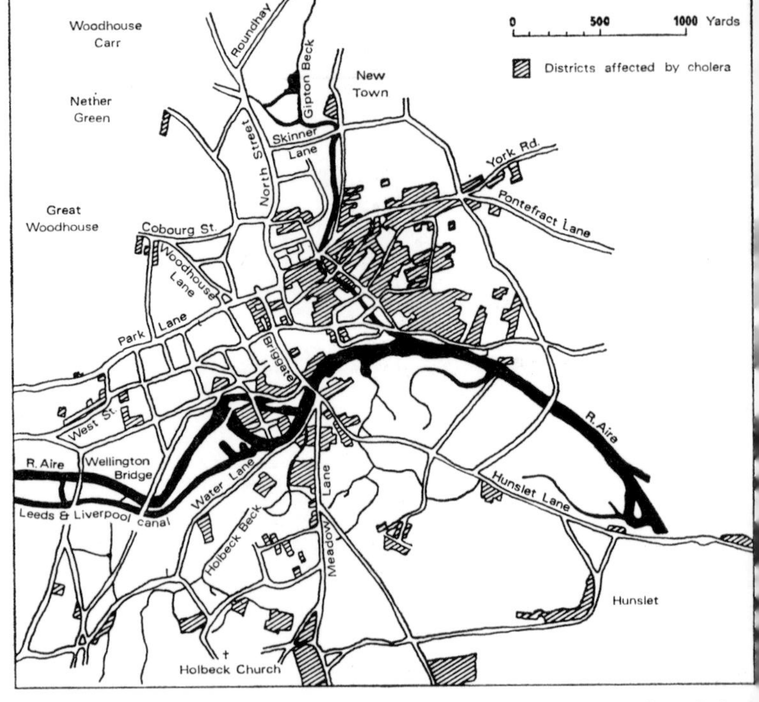

Fig 48 Robert Baker's 'cholera plan' of Leeds, 1833 (*adapted from Gilbert 1958*)

early July to the last week in August during which 400 people died. The West Country received the epidemic in mid-July. Cornwall had a death roll of just over 300, Somerset rather less, but in Devon cholera took off nearly 2,000 people (Fig 49).

Dr Thomas Shapter's[23] first-hand account of the epidemic in Exeter affords a vivid picture of the ravages of the disease and its effect upon the life of that city:

This inadequate water supply combined with the deficiency of drainage, is of itself sufficient evidence, that the necessary accommodation for the daily usages of the population must have been very limited . . . they speak of dwellings occupied by from five to fifteen families huddled together in dirty rooms with every offensive

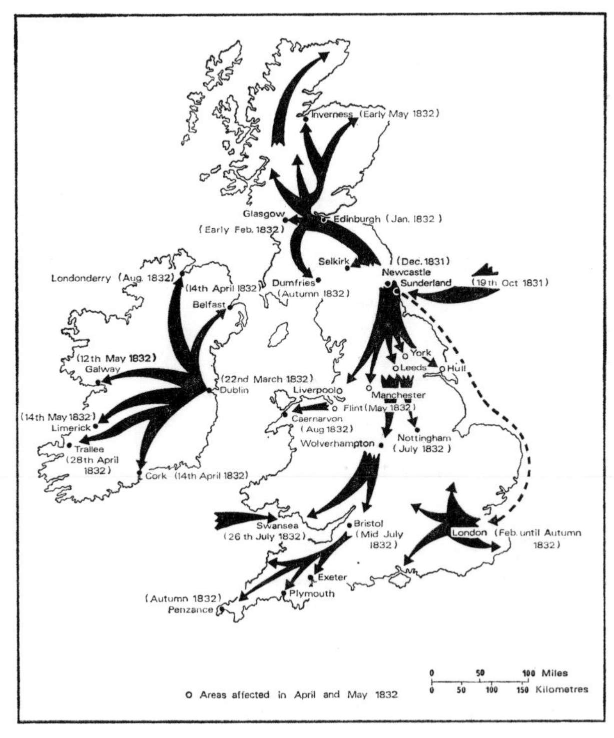

Fig 49 Progress of the 1831–2 cholera epidemic through the British Isles (*based on texts of Creighton 1965 and Longmate 1966*)

accompaniment; slaughter-houses in the Butcher Row, with their putrid heaps of offal; of pigs in large numbers kept throughout the city . . . poultry kept in confined cellars and outhouses; of dung-heaps everywhere . . .

Such conditions prevailed in the lower-lying and neglected portions of Exeter. And:

Amid this desolation, the profligacy and drunkenness of the lower orders increased to such an alarming extent as to become a matter of public remark and censure.

The epidemic began in the second half of July but was almost over by the autumn. In that time, however, there were over 400 deaths in the city out of a population of 28,242. The story of cholera in Exeter could apply equally to several other towns in the early nineteenth century; it was, in microcosm, the story of cholera in Britain.

Cholera reached Flint in North Wales early in May 1832 and spread quickly to Holywell (49 deaths), and widely into North Wales by late July. It occurred in Caernarvon during August and the early part of September causing 30 deaths. 'The terror and dismay which reigned in this town during the prevalence of the disease can never be forgotten by its inhabitants' (*Carnarvon Herald*, 22 September 1832). In South Wales the towns of the coal-field experienced severe outbreaks of the disease, both in 1832 and in the subsequent epidemic years of 1849, 1854, and 1866. It appeared in Newport on 24 June 1832, and independently in Swansea on 26 July when the *Mary Ann* called at the port with two of her crew dying from the disease. The disease continued thence to Llanelly, Neath, Haverfordwest, Merthyr Tydfil, and Builth. Merthyr Tydfil had 160 deaths from cholera during the 1832 epidemic and Swansea had 152 deaths.[24]

In Ireland the first undisputed cases of cholera were reported in Dublin in March 1832. Thereafter it spread throughout the whole country and caused great terror among the populace (Fig 50). Ireland was not to be free until well into 1834 by which time she had suffered 25,378 deaths out of a population of 7,800,000. In England and Wales the death roll was 21,882 out of a total population of almost 14 million, Scotland lost 9,592 of its 2,300,000 people.

The Cholera Acts were rushed through an agitated British Parliament but cholera came again and again. For ten years Britain remained immune from the disease, then the assault was renewed with a brief but very severe epidemic in the south of

Scotland in mid-winter 1848–9 (Fig 51). It arrived at Leith and Edinburgh at the beginning of October and reached Glasgow by December, casting a blight upon Hogmanay celebrations. Edinburgh suffered 450 deaths, Glasgow 3,800, and Scotland, as a whole, 7,000 to 8,000.

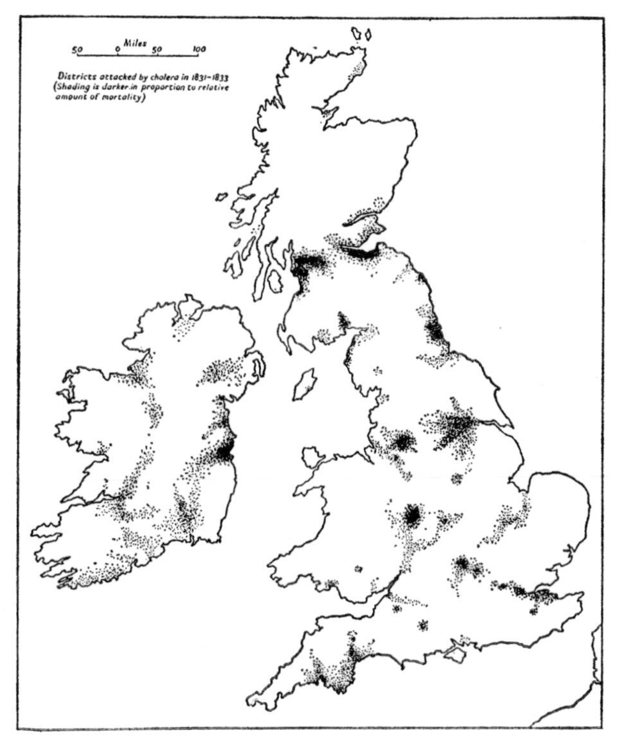

Fig 50 Areas of the British Isles affected by cholera, 1831–3 (*Petermann 1852*)

The epidemic in Scotland was over before the disease really began in England. It started in London, probably reintroduced into the country by a seaman, John Harrold, from Hamburg. It spread first south of the River Thames in the Lambeth and Southwark areas. During 1849 it raged practically the length and breadth of England causing 52,293 deaths. The East Riding of Yorkshire, Lancashire, Northumberland and Durham, Stafford-

Fig 51 Principal locations in the British Isles affected by cholera, 1848–9 (*based on text of Creighton 1965, et al*)

shire, Devon, and South Wales suffered severely. In South Wales the distress among the people of Merthyr Tydfil attracted particular attention. At the time this town, with about 50,000 people was the largest in Wales and, like Aberdare, Rhymney, Tredegar, and Ebbw Vale along the northern edge of the coalfield, had experienced a mushroom growth as an iron-smelting centre. Rows of cottages had been erected in haste to house the vast influx of workers. The result was a huge labour camp. The town lacked practically every amenity and conditions for men and women were sordid. 'The body and habits of the people [of Merthyr Tydfil] are almost as dirty as the town and houses in which they swarm; the people are savage in their manner, and mimic the repulsive rudeness of those in authority over them.' Such was the comment of a Government Commissioner in 1847. The town was without question, unsalubrious and typified the worst excesses of mid-Victorian industrial squalor.

Cholera had broken out on the coast at Cardiff in May 1849 and caused almost 400 deaths. In subsequent months it made enormous ravages in the mining valleys of South Wales. It broke out in places as far apart as Newport, Swansea, and in Holyhead in North Wales, but it was Merthyr Tydfil, 20 miles north of Cardiff, which suffered most. Cholera appeared in the town on 21 May and from then until well into November the epidemic raged and 'hardly a house escaped without feeling the lash of this scourge'. By which time 1,400 of the citizens of Merthyr had died.

It is not strictly part of the 'period-picture' to which this Chapter is devoted, but mention must be made to the 1853–4 visitation of cholera, since it was this epidemic, beginning on Wearside, which stimulated the second edition of John Snow's essay 'On the mode of communication of Cholera' in 1855. This edition makes special mention of the Soho (London) epidemic of cholera and includes a map of the distribution of cholera deaths in the Broad Street (now called Broadwick Street, Soho) district in 1854 (Fig 52). The Rev Henry Whitehead, Vicar of St Luke's, Berwick Street, in Soho, wrote of the epidemic as 'limited in its extent, brief in its duration, continually on the wane from the moment of its appearance'. The 'cholera field' as Snow called it, had its centre at the pump in Broad Street, near Golden Square, and the 'field' was bounded roughly by Great Marlborough Street, Dean Street, Brewer Street, and King Street. Within this small area over 500 people died from cholera in the ten days from 1 to 10 September 1854. Snow demonstrated that most of these deaths occurred among those who consumed water from a pump in Broad Street whereas those living in the same neighbourhood but using other water supplies escaped. He insisted that an engineer examine the well below the pump and established that the water

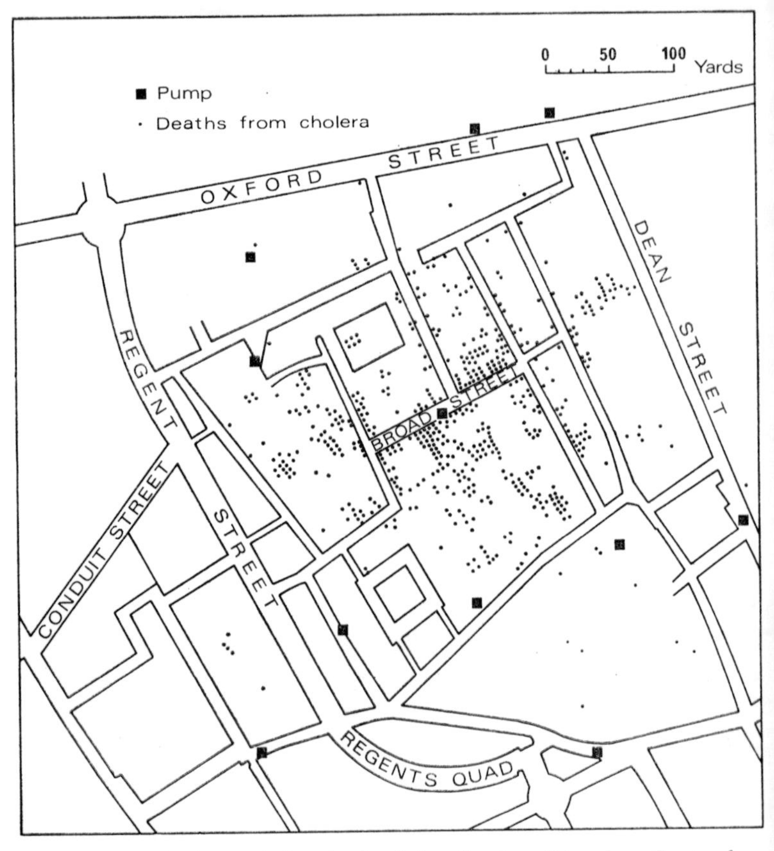

Fig 52 Deaths from cholera in the Soho district of London, September 1854 (*adapted from Snow 1855*)

had become contaminated by seepage from a leaking cesspool or drain. It was arranged to have the pump handle removed. The numbers of new and subsequently fatal cases during the critical fortnight of August–September 1854, as given by Snow, suggest that the outbreak was already limiting itself and that the value of removing the pump handle was largely symbolic. The weight of positive evidence of the source of the disease was that of the distribution map, aided by such sidelights as the case of the woman who lived at a distance but had her water brought from the Broad Street pump (because she liked the taste) and caught cholera (plate p 203).

Pulmonary tuberculosis (phthisis) was the most widespread and persistently deadly disease of the nineteenth century. Under the name 'consumption' it had appeared in the Bills of Mortality of the seventeenth century and was certainly prevalent in the eighteenth century. However it was in the nineteenth century in association with overcrowding in factories and slums that it thrived. It was a dreaded disease and not well understood. Diagnostic precision was lacking and it was frequently confused with other diseases. Yet the first analysis by the Registrar-General[25] for England and Wales in 1839 revealed that identifiable consumption alone accounted for 17·6 per cent of all deaths.

On the basis of information in the London Bills of Mortality Brownlea[26] shows that the proportion of all death dues to tuberculosis in London rose throughout the eighteenth century, reached a peak in the fifty years between 1780 and 1830, and fell steadily thereafter. Other researchers put the effective turning-point in tuberculosis mortality near the half century. Be that as it may, it is clear that tuberculosis was overwhelmingly the most important single cause of death in the 1830s. The disease had a long epidemic cycle.

Tuberculosis is caused by the *Mycobacterium tuberculosis*, discovered by Robert Koch in 1882. The two types, human and bovine, can cause disease in man. The disease may be contracted either by drinking infected milk from tubercular cows or by inhaling the bacilli when some person with the disease coughs or spits. Bovine tuberculosis was particularly common in nineteenth-century Britain and the relationship with cows' milk was not known. The conditions under which cows were then kept made it highly probable that every drop of milk was heavily charged with the organisms responsible for bovine tuberculosis.

Unsightly scrofulous glands were a frequent sight in people

suffering from tuberculosis. The Rev Patrick Brontë, father of Charlotte, Emily, and Anne, was so afflicted. The disease thrived in deprived bodies; its allies were undernourishment, debilitation, unventilated homes and working accommodation, and squalor. Such circumstances were ever-present among certain of the working population[27] in early nineteenth-century Britain, particularly among the vast army of spinners and handloom weavers in the cotton and woollen industries, the frame-work knitters of Nottinghamshire and Leicestershire, the silk workers of Coventry, the nail-makers of Birmingham, the lace-workers, shoe-makers, together with general labourers and porters.

> The staple diet of the manufacturing population is potatoes and wheaten bread, washed down by tea or coffee. Milk is but little used. Meal is consumed to some extent, either baked into cakes, or boiled up with water, making a porridge at once nutritious, easy of digestion and easily cooked. Animal food forms a very small part of their diet and that which is eaten is often of an inferior quality. In the class of fine spinners and others, whose wages are very liberal, flesh meat is frequently added to their meals. Fish is bought to some extent though by no means largely; and even this is not till it has undergone slight decomposition, having been first exposed in the markets, and being unsaleable, is then hawked about the back streets and alleys, where it is disposed of for a mere trifle. Herrings are eaten not unusually; and though giving a relish to their otherwise tasteless food, are not very well fitted for their use. The process of salting which hardens the animal fibre, renders it difficult of digestion, dissolving slowly, and their stomachs do not possess the most active or energetic character. Eggs, too, form some portion of the operatives' diet. The staple, however, is tea and bread. Little trouble is required in preparing them for use; and this circumstance joined to the want of proper domestic arrangements, favours their extensive use among a class so improvident and careless as the operative manufacturers.[28]

Burnet[29] points out that town life had important effects on food habits. It necessarily meant a greater dependence on professional services of bakers, brewers, and food retailers, partly because living conditions were generally overcrowded and ill-equipped for the practice of culinary arts, and because many wives worked at factory or domestic trades, and had little time or energy for cooking. 'The kind of food which most commended itself was, therefore, that which needed least preparation, was tasty, and if possible, hot, and for these reasons bought bread, potatoes boiled or roasted in their jackets, and bacon which could be fried in a matter of minutes, became the mainstays of urban diets. Tea was also essential because it gave warmth and comfort to cold, monotonous food.'

The first half of the nineteenth century was a time of unprecedented malnutrition. It was a hungry half century. The diet of the majority of town dwellers was at best stodgy and monotonous, at worst hopelessly deficient in quality and nutriment. Agricultural labourers fared little better, although those in the northern counties of England and in Scotland seem to have been more fortunate than those in the south. In the north oatmeal was made palatable by the addition of milk which was rarely available to the southern labourer who had no cow pasture of his own. Potatoes were a popular item of diet in Scotland, Cumberland, Westmorland and in the northernmost counties generally 'the children of our gentry prefer potatoes to bread' but they were regarded with considerable scorn by labourers in the south of England. In many respects it might be said that the agricultural labourer and his family lived near to or on the verge of starvation.

Living conditions in the towns, have already been referred to (*see* pp 159 ff). The description by Dickens of 'the neighbourhood beyond Dockhead in the Borough of Southwark (London)' is as grim as anything Chadwick or Engels ever wrote:

> . . . a maze of close, narrow and muddy streets . . . tottering house fronts, projecting over the pavement, dismantled walls that seem to totter as he passes, chimneys half crushed half hesitating to fall . . . Crazy wooden galleries common to the backs of half-a-dozen houses, with holes from which to look upon the slime beneath; windows broken and patched, with poles thrust out, on which to dry the linen that is never there; rooms so small, so filthy and squalor which they shelter; wooden chambers thrusting themselves out above the mud, and threatening to fall in—as some have done; dirt-besmeared

walls and decaying foundations; every repulsive lineament of poverty, every loathsome indication of filth, rot and garbage.[30]

It was under these conditions that tuberculosis thrived. Tubercular infection was rife among the ill-fed poor and a great many cripples owed their misfortune to this. It was considered part of life, apparently inevitable and accepted mutely. Hobson[31] suggests that much of the dramatic poetry and drama written in the nineteenth century would never have been created but for the stimulus of tuberculosis 'probably the diminished physical vigour brought about by disease increases the urge to mental activity'. Certainly the ideal of feminine beauty at that time was a languorous pale creature, lying upon her couch, dressed in white flimsy drapery. How far this was because of the prevalence of tuberculosis, or the cult among the 'ladies' of keeping themselves out of the sunlight so as to appear 'genteel' is difficult to assess.

The nineteenth century set no great premium on infant life. Child deaths were inevitably most frequent in the dreadful environment of the slums of towns. Of the one-roomed houses of the slum backlands of the industrial towns it is said that one in every five of all children born in them never saw the end of their first year and of those who so prematurely died one-third were never seen by a doctor in the course of their illness. One out of every five children who survived infancy died before they were fifteen. Ferguson[32] tells that 'in the new industrial world children were apt to be a nuisance, a drag on their parents, until they came to be old enough to earn a wage' and refers to infanticide and near infanticide and to the administration of opiates to the young in the Lancashire cotton towns (a dose of 'quietness' was given to the child to prevent it being troublesome while the mother was out to work), with disastrous effects. Infant mortality was excessive not only in the towns but also in the countryside where 'herded together in cottages which by their imperfect arrangements, violated every sanitary law, generated all kinds of disease . . . (men were) compelled by insufficient wages to expose their wives to the degradation of field labour, and to send their children to work as soon as they could crawl'.[33]

Children suffered severely from bronchitis, pneumonia, summer diarrhoea, the common infectious diseases of childhood and rickets. Diarrhoea was associated with general and domestic uncleanliness just as diseases of the respiratory organs were due to conditions of the atmosphere and typhus to overcrowding and vermin. The common infectious diseases such as scarlet fever, measles, and whooping cough were of a relatively mild character; it was the complication rather than the original disease which caused death and disability. For example, death from measles was often due to a secondary broncho-pneumonia. Rickets was a national scourge, with thousands of infants suffering from bent limbs and curved spines. Faulty diets were to blame, not least the proprietary baby foods. The latter were preponderantly farinaceous and consisted of flour starch, malted flour, and similar materials. They were deficient in protein, in fat, and in most of the vitamins. Proprietary baby foods were responsible for appalling amounts of malnourishment and sickness.

Notes to this chapter are on pp 255–7.

12

Late Victorian Times

The main theme in the history of the Victorian era was that of economic and social change and the cultural response to it. Consequently the late Victorian period contrasted markedly with early Victorian times. By the end of the nineteenth century Britain had lost her industrial and commercial supremacy to such countries as Germany and the USA. She was no longer 'the workshop of the world'. The pace of her economic growth had slackened and readjustment was taking place. The Corn Laws had been repealed, cheap North American and Russian wheat was flooding the market and those branches of agriculture concerned with the growing of cereals were depressed. Yet the population of the country continued to increase. It doubled within the two-thirds of a century of Queen Victoria's reign. The 16·3 millions in 1831 rose to 23·1 millions by 1861 and 33 millions by 1891. By the 1890s, 70 per cent of the population dwelt in towns. Birth rates remained high and death rates continued to decline, although after about 1880 there were indications that the birth rate was beginning to fall (Fig 53).

The late Victorian era experienced the last of the cholera epidemics (1866), an epidemic of smallpox (1871), and a pandemic of influenza (1889). It also embraced the 'ages' of Chadwick and Simon.

Edwin Chadwick (1800–90),[1] author of the *Report on the Sanitary Condition of the Labouring Population of Great Britain (1842)* (*see* p 159) was Secretary to the Poor Law Commission. The statement on the general condition of the dwellings of the

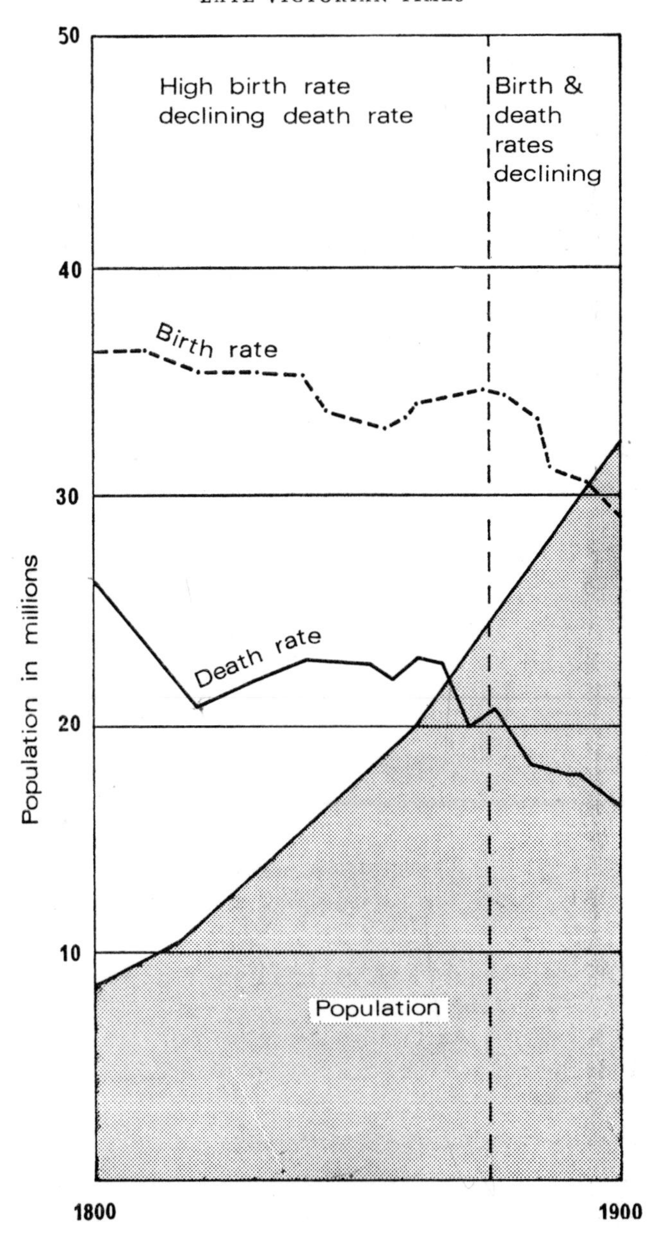

Fig. 53 Birth rates, death rates, and population totals in Britain, 1800–1900

labouring classes or of workmen's lodging houses, the sanitation of their homes and places of work, and the domestic habits affecting the health of the labouring classes contained in his *Report* demonstrated the need for national action and was instrumental in forcing questions of public health into politics. Chadwick waged war against dirt and disease and has been called 'Father of the Sanitary Idea'. The Public Health Act, passed in 1848, was very much linked with him, and with his co-workers Dr Southward Smith, Dr Neil Arnott, Dr J. P. Kay, and Dr John Snow.

John Simon (1816–1904), Medical Officer of Health for the City of London and later at the office of the Privy Council, was first public health pioneer after Chadwick and drew heavily on his work. He promoted several inquiries into the incidence and mortality of diseases in different areas of the country which had an enormous influence in promoting sanitary reform. At this same time William Farr was Compiler of Abstracts at the General Register Office. It was he who analysed the available data relating to life and death in Britain which was used to such effect by both Chadwick and Simon in their crusade against insanitary conditions.

The late nineteenth century was a time when the role of micro-organisms in the causation of disease had yet to be demonstrated. Many diseases were still thought of as being initiated by the diffusion of gaseous material from poisons in the soil, usually of putrefactive origin, from decomposing animal and vegetable substances, from damp and filth, and from close and over-crowded dwellings. Odours and emanations (miasmas) were considered the responsible agents.

Until late in the nineteenth century medical men knew little more than had their Greek forbears about actual causes of plague, fever, and pestilential scourges. By the end of Pasteur's career, near the end of the century the rudimentary germ theory had been proved and was no longer seriously contested; the patterns of many infective diseases were understood; methods had been devised for preventing or for combating some of the most serious infections; conditions under which surgical procedures were carried out had been revolutionised; and the sciences of bacteriology and of preventive medicine had been launched.

Hieronymus Fracastorius (*see* p 115), Marcus Antonius von Plenaz and others had approached earlier the correct explanation of the nature and causation of disease but they were incapable of providing the necessary proof. The development of the microscope in the seventeenth century made it possible to demonstrate the existence of living unicellular organisms by actual observation, but it was Louis Pasteur (1822–95) and Robert Koch (1843–1920) who provided undisputable proof of the existence of germs, of their modes of reproduction, and of their specificity in causing disease. This work culminated in preventive inoculation against hydrophobia in 1885. Pasteur prophesied that microbes could produce disease, but it was Robert Koch who provided the proof. Koch isolated the tubercle bacillus (*Mycobacterium tuberculosis*) in 1882 and identified it as the specific organism of pulmonary tuberculosis (consumption, phthisis). By the end of the century the organisms causing diphtheria, lobar pneumonia, erysipelas, Malta fever, cerebro-spinal meningitis, tetanus, plague, botulism, cholera, dysentery and wound infection had been identified, isolated, and proved to be responsible for the diseases with which they were associated. Equally important was knowledge of the usual sources of the organisms and the routes by which they travelled. Not until the closing years of the nineteenth century was the part played by insects (ticks, flies, fleas, etc) in the transmission of disease discovered. Armed with such knowledge, ways and means for blocking the routes and preventing further infections were made possible.

Influenza had been present in Britain in fairly severe epidemic form during the nineteenth century (1803, 1831, 1833, 1837, 1847–8), but Greenwood,[2] writing in 1935 said: 'In 1889 this country had been free from pandemic influenza for more years than in any previous epoch since the middle of the seventeenth century.' That year, however, saw the beginning of a pandemic which was to last until 1894, with recurrences in 1895, 1900, and 1908, until it culminated once again in the great pandemic of 1918–19. Though its source was in doubt the pandemic of 1889 seemed to have originated in south-west Siberia in what was then Russian Turkestan (ie Soviet Middle Asia), and to have spread rapidly westwards to western Europe, reaching Britain in January 1890. By February it had spread throughout the whole of the British Isles. It was a mild form of influenza and did not cause the high mortalities experienced in many countries on the Continent. This epidemic of 'Russian influenza' was followed by three more waves, which had 'peaks' in May 1891, January 1892, and December 1893. The first (1890), was nationwide, the second (which began in Hull and on the Welsh Border) and third (first reported in the west of Cornwall and in the east of Scotland) were more localised and more desultory or prolonged. The revival of the epidemic in three successive seasons made the late nineteenth century invasion of the disease unique since no similar sequence had been recorded in the previous history of the disease in Britain.

Influenza is a virus disease of which there are three known antigenic types, A, B, and C. Influenza viruses A and B have been implicated in several major epidemics. Sporadic cases or localised outbreaks of influenza C are the rule. Influenza virus is transmitted by direct contact and droplet infection. The virus of the 1889 epidemic was due to a sudden mutation of the virus type A rendering it slightly different from and more virulent than the strain which had previously been present in Britain. Such a mutation enabled the virus to produce infection among individuals who would have escaped prior to the alteration. Having acquired no immunity, people were only able to put up the feeblest of resistance to the epidemic in 1889.

Influenza is a winter disease as a rule (Fig 54) and it requires relatively low temperatures for its propagation. It also needs a low level of immunity in large numbers of the population and the presence of a suitable type of virus at the time to enable the epidemic to occur. Evidently these environmental conditions were present in combination in Britain in 1889.

Though not in correct chronological sequence it is convenient, for the sake of completeness, to make brief mention in this Chapter of the 1918–19 pandemic of influenza. Taken in its global con-

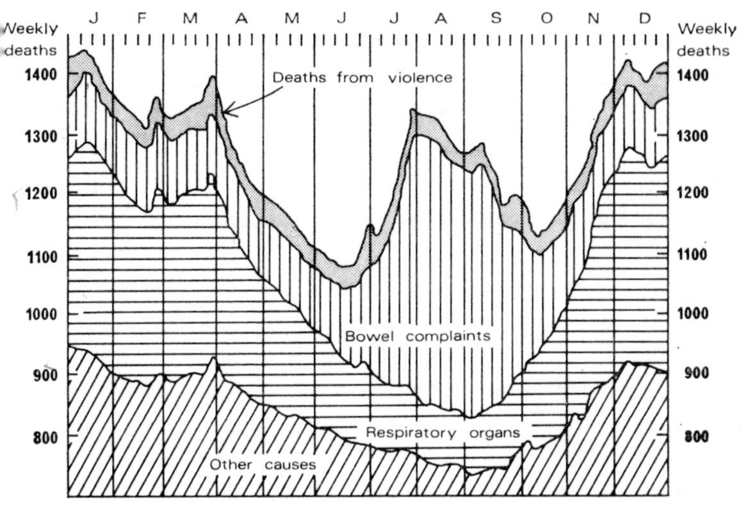

Fig 54 Seasonal trends of deaths in London, 1845–74 (*after Buchan and Mitchell 1875*)

text, this was one of the most destructive pandemics in history and ranks with the Great Pestilence or Black Death as one of the severest holocausts of disease ever encountered. Though commonly thought of as a disease of the colder part of the year the first wave of prevalence in Britain came in May, June, and July 1918. The second wave came in October and November 1918 and was much more severe than the first. It tended to kill the young rather than the elderly. A third and final wave occurred in February 1919. There were all told about 150,000 influenza deaths in England and Wales during the epidemic; in Scotland there were 3,776 deaths in Glasgow alone.

Pneumonia was frequently associated with influenza during the 1918–19 pandemic. Indeed there might have been a concurrent epidemic of pneumonia in Britain at the time. As a disease in its own right pneumonia in 1900 was fourth major cause of death in Britain after heart disease, tuberculosis, and bronchitis. For any person to be in poor health was an invitation to pneumonia. Pneumonia, inflammation of the lungs, usually results from *Pneumococci*, or other organisms already present in the throat or inhaled from outside sources, spreading down and multiplying widely in the substance of the lung. As with influenza it is most common in cold months. Evidently at the time of the influenza pandemic the whole population was also susceptible to pneumonia. There were, in addition, viral pneumonias and a haemolytic streptococcal form present as complications of the influenza. Young children were particularly vulnerable and pneumonia, after diarrhoea, was probably the second most important cause of death of children under five.

Under-nourishment was a possible contributory cause of the influenza and pneumonia epidemics of 1918–19. Food shortages and rationing were fairly severe during World War I following the 1917 German U-boat campaign. There was not the same attention given by the Government of the day to the provision of bread of high nutritive value, to milk supplement for expectant and nursing mothers and all children up to the age of 15 years, and to fortifying margarine with vitamins A and D, as during World War II.

Bronchitis the name given in the early nineteenth century by Charles Badham[3] to 'the more chronic pectoral (chest) complaints, especially those of people advanced in life' flourished in Britain at the end of the Victorian era particularly in the damp and smoky atmospheres of the new industrial areas. It was a disease of the industrial towns of Lancashire, the West Riding of Yorkshire, the Midlands, the Black Country, South Wales and Clydeside. Bronchitis is essentially inflammation of the mucous membrane of the bronchi. A cold environment predisposes to the disease as it does to pneumonia and other respiratory infections and inflammation is liable to be set up by a sudden change from warm to cold air. Air polluted by toxic substances, inspired with a humid atmosphere, also irritates the respiratory passages. Acute bronchitis may be precipitated by excessive atmospheric pollution; chronic bronchitis refers to long standing inflammation of the bronchi. Whatever the conditions, the possible causes (including viral or bacterial infection) were not known at the turn of the century—indeed they are elusive even in the second half of the twentieth century.

In the 1830s scarlet fever took over the mortality trail which until then had been blazed by cholera, influenza, and smallpox. For a long time it was impossible to differentiate it from diphtheria,[4] because both diseases had, in common, the symptom of a sore throat. Typically, however, scarlet fever had, in addition to the sore throat, the rash; in diphtheria, the sore throat was sometimes associated with paralysis (especially of the soft palate) and obstruction of the larynx, popularly known as 'croup'. During the middle years of the nineteenth century scarlet fever (scarlatina)

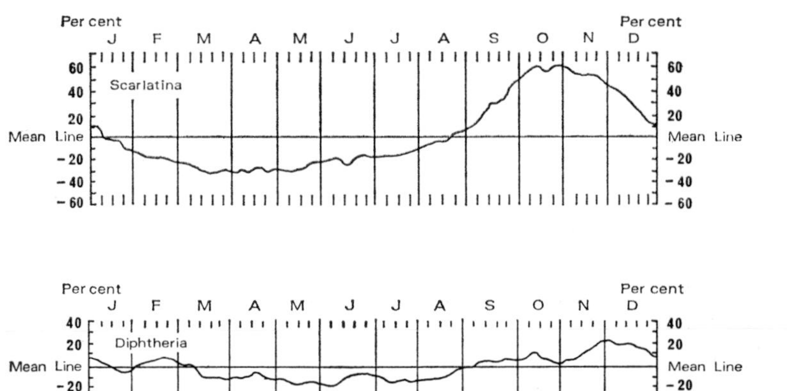

Fig 55 Seasonal trends in deaths from scarlatina and diphtheria in London, 1845–74 (*after Buchan and Mitchell 1875*)

was the leading cause of death among the infectious diseases of childhood and during 1863—the year with the highest mortality—the death rate for children under fifteen was 3,966 per million. There were over 30,000 deaths for scarlet fever in 1863 (Fig 55). Scarlet fever is the result of infection with *Streptococcus haemolyticus* which may remain quiescent for a long time in dust or clothing but which reproduces rapidly in organic substances, notably milk. From its peak in the 1860s, scarlet fever showed a marked decline which continued to the turn of the century and, in fact, to the present day. Why it changed from being a dreaded children's disease to a mild illness affecting older age groups—a change which long preceded the use of antibiotics—is still unknown. It may have been due to a reduction in the virulence of the various types of beta-haemolytic streptococci.

At first diphtheria was overshadowed by scarlet fever, but a wave of prevalence started in the mid-nineteenth century which brought it very much to the fore. The outbreak of 1856–9 was part of a sudden uprising of the disease throughout the world. Creighton[5] tells of fatalities from 'inflammation of the throat', 'putrid sore throat', 'malignant sore throat', 'disease in the throat' in Cornwall (Launceston, Liskeard, Truro), Lincolnshire (Spalding), Kent (Ash), Oxfordshire (Thame), Essex (Billericay, Maldon), and Derbyshire (Chesterfield). The distribution of the epidemic suggested that diphtheria might have been a country disease since agricultural counties appeared to have had somewhat more than their usual share of an infective mortality compared with the industrial centres. After the initial outburst the apparent rural preference of the disease disappeared but a high death rate of about 800 per million population persisted until about 1900.

The main causes of death at the end of the Victorian era may be summarised as follows:

	%	
Heart diseases	12·8	(largely old age)
Tuberculosis	10·4	
Bronchitis	9·2	
Pneumonia	7·5	
Vascular lesions of nervous system	7·0	(largely old age)
Cancer	4·5	
Accidents	3·1	
Infant mortality (measles, whooping cough, diphtheria, scarlet fever)	23·5	
All others	22·5	
	100·0	

Figs 56 to 59 show, for England and Wales only, the area distribution of average death rates for All Causes, Zymotic diseases, Phthisis and Infant Mortality respectively in 1901.

Living conditions in the late nineteenth century improved in ways which quite certainly influenced the course of infectious diseases. Accumulated knowledge of the conditions of life among the industrial poor and of the part which environmental conditions played in predisposing to the disease stimulated sanitary endeavour. Better housing, cleanliness, ventilation, disinfection,

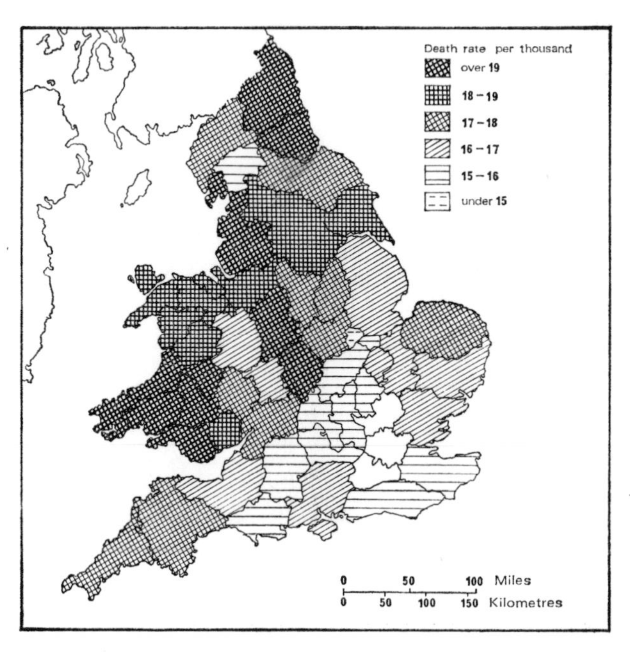

Fig 56 England and Wales: distribution of average death rates from 'All Causes', 1901 (*adapted from Bartholomew 1902*)

control of nuisances, improved water supplies and improved refuse and sewage disposal were some of the measures which were applied successfully although with no significant improvement in mortality rates until after 1871.[6] Preventive measures did not rest on knowledge of the mechanism of infection but derived empirically from noting the association between sickness and bad living conditions. The effectiveness of the measures was later explained rationally by the science of bacteriology.

Against no disease did the sanitary improvements of the late nineteenth century win greater triumphs than typhus, unassisted by bacteriological research or by knowledge of the bionomics of vermin. There were sharp falls in the death rates and probably in the incidence of typhus and enteric fevers (typhoid and paratyphoid). Cholera had virtually disappeared, so had smallpox after

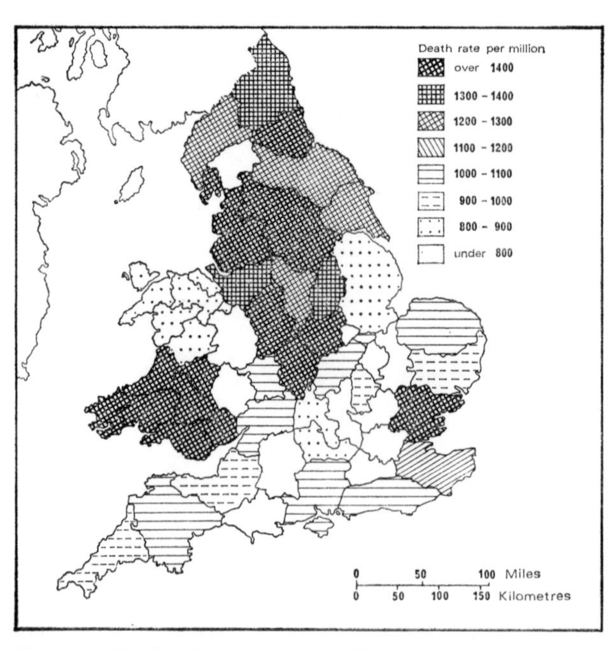

Fig 57 England and Wales: distribution of average death rates from zymotic (ie infectious and contagious) diseases, 1901 (*adapted from Bartholomew 1902*)

the implementation of the Vaccination Act of 1861 (Fig 57). The risk of epidemic and crowd diseases fostered by poor sanitation and overcrowding was decreasing but in the towns there was a high risk of degenerative disease resulting from harder wear and tear of factory employment and urban discomfort. The Census of 1901 disclosed that about 16 per cent of the population of London lived in overcrowded conditions. Higher proportions existed in Dudley (17 per cent), Devonport (17 per cent), Plymouth (20 per cent), Sunderland (30 per cent), Newcastle-upon-Tyne (30 per cent), Tynemouth (31 per cent), South Shields (32 per cent), and Gateshead (34 per cent).

Tuberculosis, the *white plague* ('where youth grows pale, and spectre thin, and dies'[7]) was still a much dreaded disease (Fig 58). Infant mortality, too, continued at a high rate (172 per 1,000

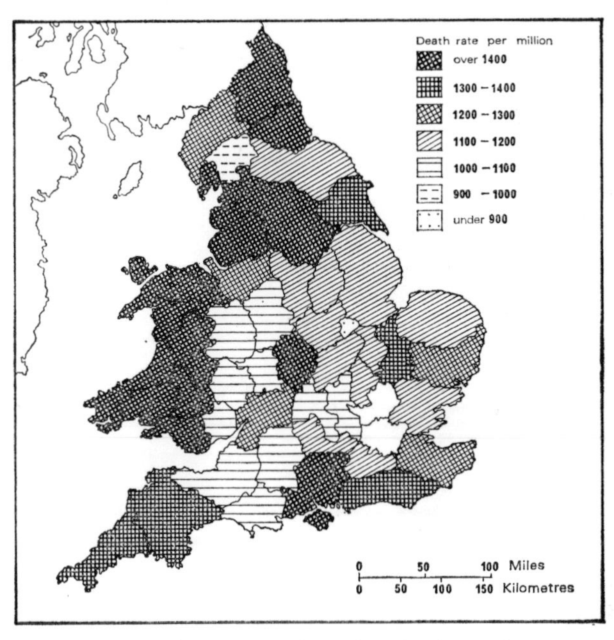

Fig 58 England and Wales: distribution of average death rates from phthisis, 1901 (*adapted from Bartholomew 1902*)

during the period 1891–1900) but there was a change in the relative importance of the fatal diseases. Scarlet fever, previously the most fatal of the common infectious diseases of childhood was, by the turn of the century, fourth after measles, whooping cough, and diphtheria (Fig 59).

The end of the Victorian era was a time of declining birth and death rates. The crude birth rate in 1901 was 29 per 1,000 and the crude death rate 17 per 1,000. Expectation of life at birth was

43·9 years during the period 1881–90 and 44·13 years in the period 1891–1900. It can probably be said that there was a gain of ten or more years in the expectation of life at birth during the course of the nineteenth century, indicative of progress towards healthier living conditions and greater facilities for combating disease rather than any material modification in inborn

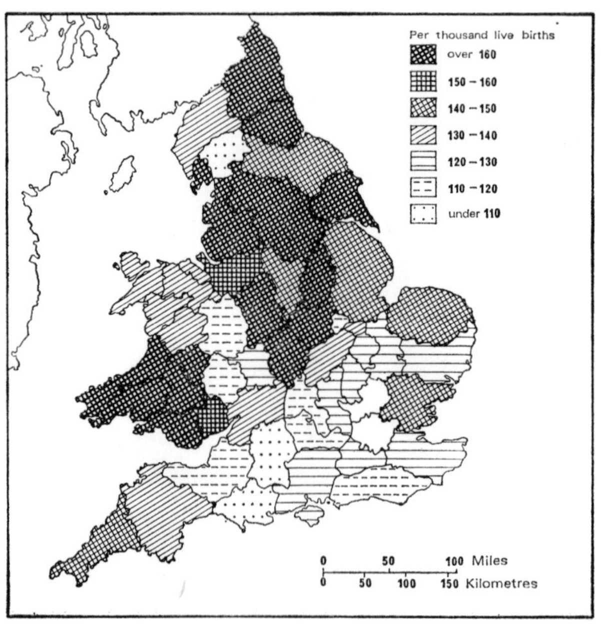

Fig 59 England and Wales: distribution of average infant mortality rates, 1901 (*adapted from Bartholomew 1902*)

characteristics of the population itself. In some respects there was a decline in the overall physique of the workers. Seemingly the forces which had produced a reduction in the death rate had not operated equally throughout every section of the population. Some sections were being missed, as was evidenced by the Recruiting Returns of the South African War. Appreciable defects in physique, health, and efficiency were disclosed. Rejections of Army recruits were as high at 60 per cent in some areas and over the whole country nearly 40 per cent. The chief grounds for rejection were bad teeth, heart affections, poor sight or hearing, and deformities. Indeed the extent of defects among volunteers for the South African War shook the public conscience. The material progress of the nineteenth century had been purchased seemingly at the cost of a general deterioration in health of the population. Drummond[8] says: 'It is no exaggeration to say that the opening of the twentieth century saw malnutrition more rife in England than it had been since the great dearths of medieval

Table 8 Expectation of life in years at birth in selected localities in Britain, 1841 and 1881-90 (*extracted in part from D. V. Glass 1964*)

Period		*London*	*Liverpool*	*Manchester*	*Surrey*	*Selected healthy districts in England*	*Glasgow*
1841	Males	35	25	24	44		
	Females	38	27		46		
1881–90	Males			29		51	35·2
	Females			33		54	44·3

and Tudor times.' Burnett[9] disagrees and considers Drummond's statement 'almost certainly untrue'. An Inter-Departmental Committee, set up to investigate the extent of the physical deterioration of the population, reported in 1904[10] and substantiated what had been amply demonstrated by Seebohm Rowntree[11] for York in 1900 and Booth[12] for London in 1902. The Committee's comprehensive survey of the possible causes of the poor physique and the ill-health of the labouring population of the towns made special mention of such factors as overcrowding, bad sanitation, alcoholism, factory conditions and ignorance. Drummond says that the survey failed to give due regard 'to what was by far the most important cause, semi-starvation due to sheer poverty' and highlights the deficiencies of white bread and the diet of children. Even so the Inter-Departmental Committee Survey did pay considerable attention to the defective diet of babies and young children. Attention was given to the rapid decline in breast-feeding, due partly to the employment of married women in industry, but more importantly, to the chronic ill-health of mothers which rendered many of them incapable of providing the necessary milk. The usual substitute in working-class homes was sweetened condensed skimmed milk, rich in sugar but almost wholly devoid of fat. When older, poor children passed to a diet consisting essentially of bread, margarine, and jam. The 1904 Inter-Departmental Committee Report found that 33 per cent of all children were undernourished in the sense that they actually went hungry. Two years earlier, a survey of Leeds had shown that in the poorest areas of the city 50 per cent of the children had marked rickets and 60 per cent had carious teeth. It comes as no surprise, therefore, to learn that at the beginning of the present century twelve-year-old boys at public or private (fee-paying) schools were, on average, 5 inches taller than those in local authority or council schools. It was this growing concern for the health of children, fortified by the complaints of teachers that hungry scholars were uneducable, that culminated in the passing of the Education Act (Provision of Meals) in 1906 and the Medical Inspection Act in 1907 (plate p 221).

The end of the Victorian era was a time of paradox. For some it was an age of affluence and well-being but for too many it was a time of pauperism and fearful mortality. Even so the modern growth of population which started in the eighteenth century continued apace into the nineteenth century (Fig 60). McKeown *et al* have discussed possible reasons for this rise of population.[13] They draw attention to the decline in mortality between 1840 and 1900 and show that 'five diseases or groups of diseases were responsible for it: tuberculosis for a little less than a half; typhus, typhoid and continued fever for about a fifth; scarlet fever for a fifth, cholera, dysentery and diarrhoea for nearly a tenth; and smallpox for a twentieth'. They further suggest that in order of relative importance the influences responsible for the reduction in mortality were '(*a*) a rising standard of living, of which the most significant feature was possibly improved diet (responsible mainly for the decline of tuberculosis and, less certainly and to a lesser extent, of typhus; (*b*) hygienic changes, particularly improved water supplies and sewage disposal, introduced by

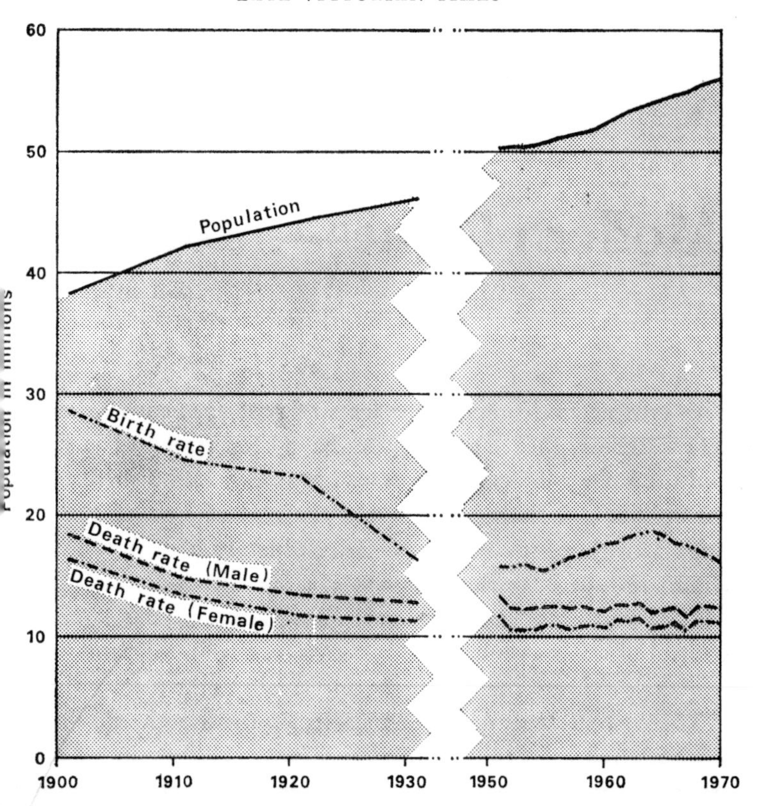

Fig 60 Birth rates, death rates, and population totals in Britain, 1900–70 (*Registrar-General's Statistical Reviews of England and Wales*)

sanitary reformers (responsible for the decline of the typhoid and cholera groups); and (*c*) a favourable trend in the relationship between infectious agent and human host (which accounted for the decline of mortality from scarlet fever and may have contributed to that from tuberculosis, typhus, and cholera). The influence of specific prevention or treatment of disease in the individual was restricted to smallpox and made little contribution to the total reduction of the death rate' (plate p 204).

Notes to this chapter are on pp 257–8.

13
Modern Times—Morbidity

The grosser environmental defects of past ages such as lack of safe water supplies and sanitation have been remedied in Britain and the killing infectious diseases have been vanquished. Indeed the mid-twentieth century marks the end of one of the most important social revolutions in British history, that is, the virtual elimination of infectious diseases as a significant factor in social life. Nevertheless an occasional breakdown in the system of control of water supplies, milk or sewage, or of public and private hygiene can bring these diseases once again to the fore.

There were typhoid epidemics in Bournemouth, Poole, and Christchurch in 1936, in Croydon in 1937, Harlow in 1963, and Aberdeen in 1964. In Croydon it was a case of contaminated sewage escaping along fissures in the surrounding chalk strata into a well from which drinking water was drawn. In Harlow and Aberdeen the disease was traced to tins of imported corned beef.[1] The problem of typhoid in Britain now is essentially that of the 'carrier'.[2] Most carriers harbour the typhoid bacillus in the gall-bladder, women being more frequently carriers than men for the reason that women are more susceptible to gall-bladder complaints. The germ can live in the gall-bladder of carriers for months or years and while carriers themselves remain apparently healthy, they can, nevertheless, go on passing the disease if they touch food with unclean hands. There are 72 types of typhoid bacilli and in the 1970s possibly as many as 500 carriers in the country. The type of bacillus excreted by each carrier is known, so that an epidemic can be traced to a known carrier once the type is recognised. This also helps to identify the possible nationality of the person and/or to identify the food concerned. There are about 200 cases of typhoid in Britain every year, half of which are associated with immigrants, visitors, or people infected on holidays abroad.

As noted (in Chapter 2) Britain is as much the creation of European emigration as is the United States. Since the turn of the century the cosmopolitan character of the population has been further enhanced by influxes of Indians, Pakistanis, and other Asian peoples, Africans, West Indians, Poles, Cypriots, Ukranians, Hungarians, Spaniards, Maltese, and Australians. There are now roughly 1·9 million foreign and Commonwealth immigrants living in Britain; one in every fifty British citizens has a dark skin. There are, in addition, one million Irish residents and an ever-changing population of students, *au pair* girls and tourists. Each year thousands of travellers enter Britain.

It is not known how much tuberculosis, smallpox, venereal disease, leprosy, and tropical disease has been brought into this country since World War II. Neither is it known how much dissemination of these and similar diseases there has been in Britain from outside sources. It is said that the incidence of tuberculosis among some immigrant colonies in Britain is three to five times greater than the national average. Immigrants with undetected tuberculosis move freely and work anywhere—even in restaurants. Outbreaks of smallpox occurred between 1961 and 1962 following the air transport of carriers from Karachi (East Pakistan) to Britain.

Such outbreaks are, however, exceptional and the concern in mid-twentieth century Britain is for the growing toll of the cancers, of cardiovascular disease, mental illness, and motor vehicle accidents. In mid-twentieth century Britain all except the latter two are degenerative diseases of middle or later life, the aetiology of which is to a large extent unknown. Such disorders have been classed as 'degenerative' but there is evidence to suggest that environmental factors play an important part in their development. In most cases the evidence is incomplete and the environmental factors have yet to be defined with precision.

When they have been defined, as for example in relation to some industrial cancers, the knowledge has enabled preventive measures to be introduced which, in some instances, have already led to the virtual elimination of these forms of disease. It seems reasonable to assume that more exact identification of the environmental factors involved may prove to be of considerable value in increasing understanding of the causation of other and more common conditions. Be that as it may, the twentieth century has brought a much healthier life and a much longer life for the people of Britain. Death rates continue to decline and the span of life enjoyed by most people is now near the biblical three score years and ten. Men and women are living longer than previously although it may be noted that the longevity of women is increasing faster than that for men. In England and Wales the expectation of life at birth for a boy is now 69 years and for a girl 75. In Scotland it is $67\frac{1}{2}$ years for a boy and $73\frac{1}{2}$ years for a girl (Table 9).

Children are also growing up earlier. Puberty is coming earlier and growth stopping sooner. The most obvious change has been a steady increase in the height of children which, since the turn of the century, has amounted to about half an inch per decade in five- to seven-year olds and about an inch per decade in ten- to fourteen-year olds. Today's five-year olds are generally two inches taller than five-year olds at the turn of the century. In eleven-year olds the difference is nearer four inches. Adults, too have been getting bigger, though not so fast as children. In girls the age of menarche (the first menstrual period) is an accurate indication of the onset of puberty. The menarcheal age has fallen from sixteen or seventeen years in the middle of the nineteenth century to around thirteen years in the late 1960s[3] (Fig 61). Many people question whether early maturity in children is desirable and wonder what the educational, social, and economic implications will be of a further lowering of the age of maturity.

Accompanying the environmental changes of the twentieth century (improved housing, nutrition, and clothing etc, but ever increasing environmental pollution) have been impressive advances in medical standards and skills. Chemotherapy and an ever-increasing range of antibiotics have further contributed to

Plate 13 'Father Thames introducing his offspring to the fair city of London' (1858)

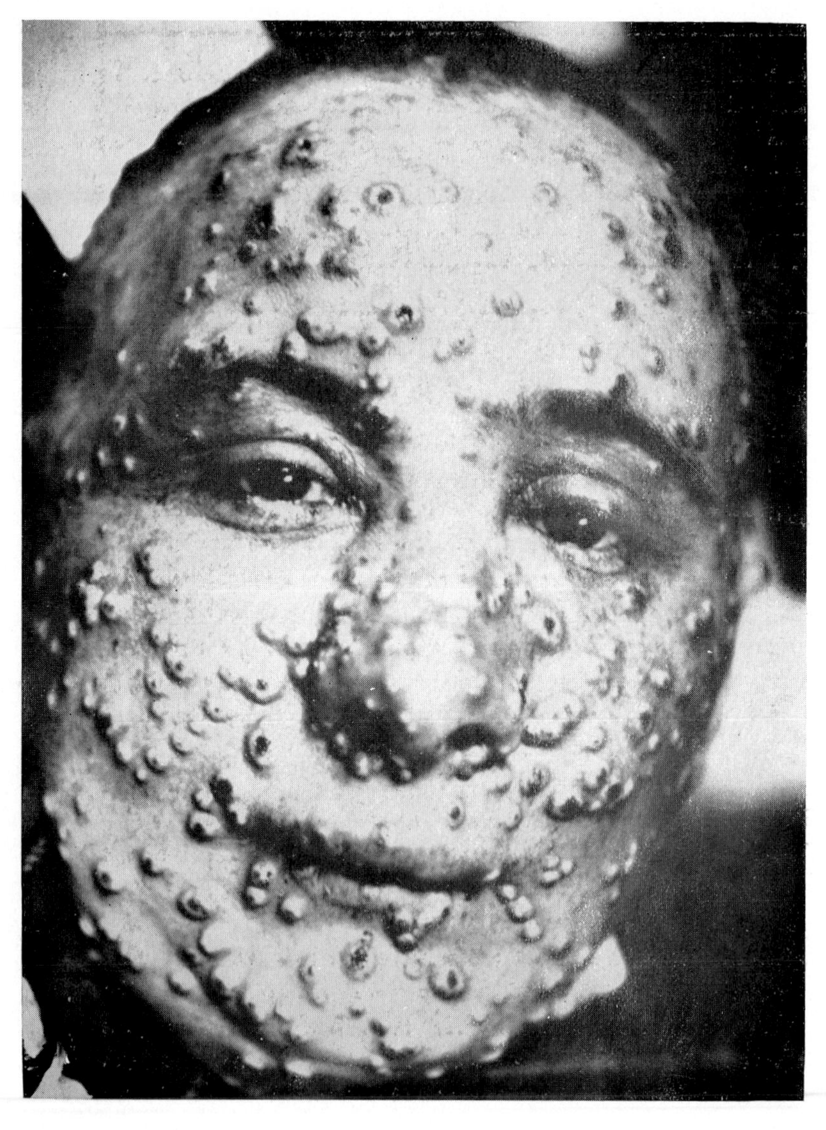

Plate 14 A victim of smallpox during the Glasgow epidemic of 1920

O *Table 9* Expectation of life in years at birth in different parts of Britain, 1841 and 1960-2 (*in part from D. V. Glass 1964*)

Period	*Male*				*Female*			
	England and Wales	*London*	*Scotland*	*Glasgow*	*England and Wales*	*London*	*Scotland*	*Glasgow*
1841	40·2	35·0			42·2	38·0		
1871–80			41·0	30·9			43·8	40·2
1881–90			43·9	35·2			46·3	44·3
1911–12	51·1	49·5			55·0	54·5		
1920–2	55·6	55·3	53·1	48·4	59·6	60·0	56·4	50·8
1930–2	58·7	59·5	56·0	51·3	62·9	64·4	59·5	55·2
1950–2	66·4	67·3	64·4	62·0	71·5	73·0	68·7	66·3
1960–2	68·1		66·2		74·0		71·9	
Early 1970's	69·0		67·5		75·0		73·5	

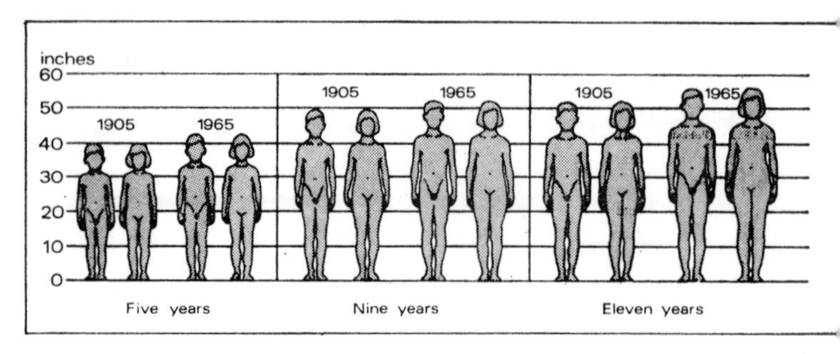

Fig 61 Heights of children, 1905–65 (*after Tanner 1968*)

changes in the pattern of disease. A century ago tuberculosis wa
the most fatal of all the diseases in Britain. The death rate wa
360 per 100,000 of the population. Fifty years ago it had droppe
to 150 per 100,000; it is now 10 for men and 3 for women pe
100,000 (Fig 62). Many factors have contributed to these results
better housing; better diets; earlier diagnosis through mas
X-rays; the prevention of infection through cows; the combine
success of antibiotics and chemotherapeutics (streptomycin
para-amino-salicylic acid (PAS), and isonaizid (INH)) in th
clinical treatment of the disease; and the BCG vaccinatio
campaign.

The recorded death rates for pneumonia rose steadily durin
the second half of the nineteenth century but have fallen steadil
during the first part of the present century (Fig 63). As wit
tuberculosis the steady decline to the 1930s can be ascribed t
rising standards of medicine and public health measures. Rate
fell steeply from the late 1930s until the mid-century since whe
they have remained relatively constant. It is reasonable t
attribute the greater part of this abrupt improvement in deat
rates to chemotherapy, namely the introduction of sulphonamide
in 1935, penicillin a few years later and the broad spectrum anti
biotics in the late 1940s and early 1950s.[4]

Chronic rheumatic heart disease, caused by rheumatic fever
has diminished as rheumatic fever has declined in severity an
incidence. Rheumatic fever itself is caused by streptococca

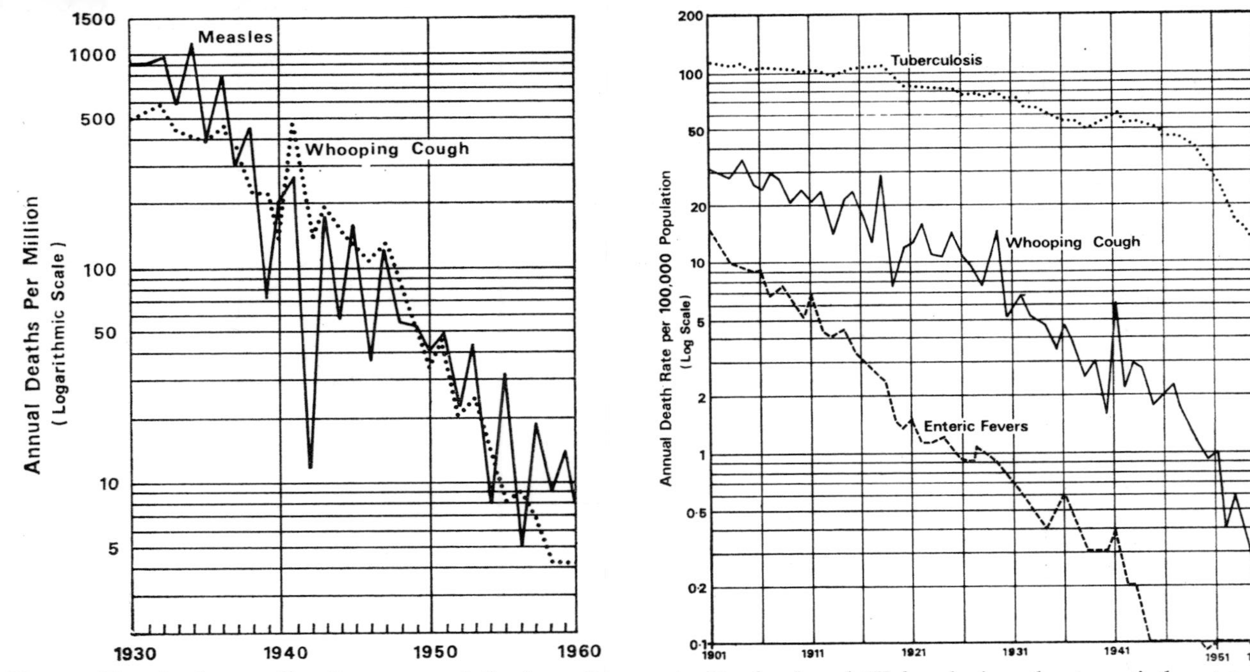

Fig 62 Trend of mortality from some infectious diseases in England and Wales during the twentieth century: (*a*) measles and whooping cough, children aged 1 to 4 years; (*b*) tuberculosis, enteric fever, and whooping cough, all ages (*after Office of Health Economics 1964*)

throat infections and the control of these infections by anti-bacterial therapy has contributed substantially to the rapid fall in mortality from rheumatic heart disease since the late 1930s.

Over the past forty years diabetes has changed from a progressive or rapidly fatal disease into a controlled chronic disorder

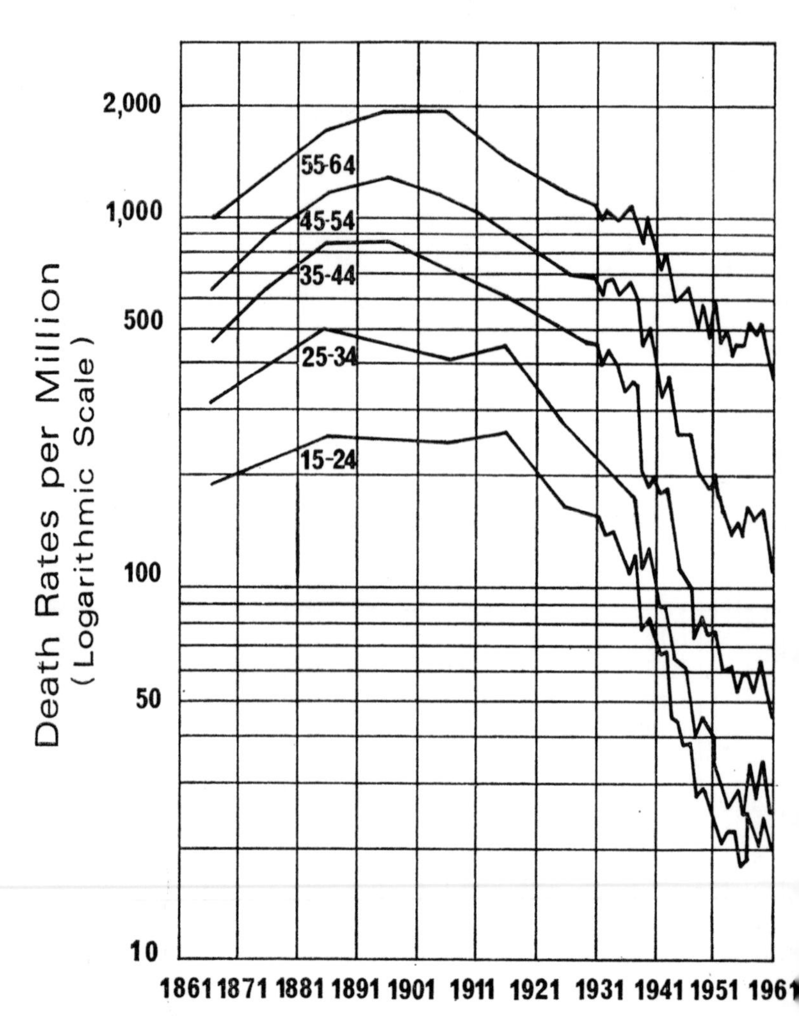

Fig. 63 Pneumonia death rates (ages 15–64) in England and Wales, 1861–1960 (*after Office of Health Economics 1963*)

with mortality confined mainly to old age. The new picture dates from the isolation of insulin in 1922.

The hygiene and sanitation of modern society indirectly

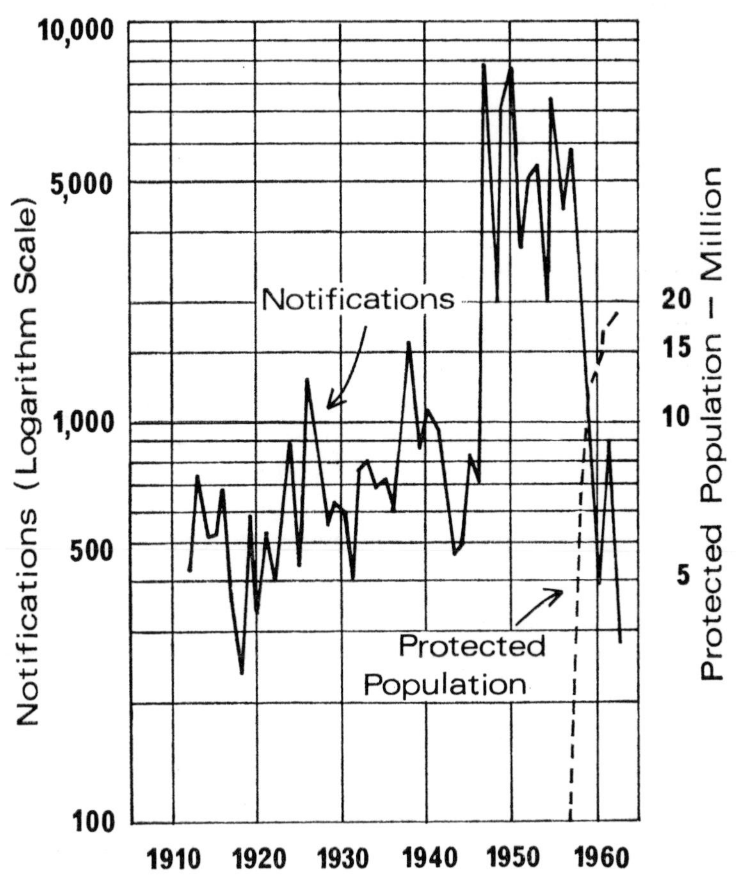

Fig 64 Poliomyelitis: annual number of cases notified in England and Wales, 1912–62 and total protected population, 1957–62 (*after Office of Health Economics 1963*)

inhibits the growth and development of immunity against poliomyelitis (Fig 64). In Britain improved sewage systems, the eradication of insanitary living conditions and greater care over infant feeding have contributed to this process. The population became vulnerable and in 1947 almost 8,000 cases were notified compared with less than 1,000 in any one year for the previous 35 years. Between 1947 and 1958 over 50,000 persons contracted the disease and deaths averaged 350 a year. The majority of the victims were young adults. Vaccination started on a small scale in 1956 and an intensive campaign was launched in 1958. Since then notifications of the disease have declined and in the 1970s poliomyelitis will probably become even more rare than diphtheria. International studies of the effects of poliomyelitis vaccine suggest that the decline can be directly attributed to the vaccination campaign.[5]

In childhood mortality diphtheria has been virtually eliminated (Fig 65). A diphtheria anti-toxin was first produced in 1890 and the diagnostic Schick test and active immunisation date from the early part of this century. In the late 1930s a national immunisation campaign was attempted. Between 1940 and 1943 nearly five million children were protected, and the results were impressive. By the 1950s immunisation against diphtheria was combined with protection against whooping cough and tetanus, and more recently poliomyelitis has been added to form a quadruple vaccine.[6]

Figs 66 and 67 show for the first half of the twentieth century the change in rates and relative importance respectively of the main causes of death in England and Wales. Nowadays the diseases which cause most distress, dislocation of human endeavour, loss of efficiency in work and of working hours are not necessarily the killing diseases (Fig 68). For this reason death rates no longer provide suitable indices of the general health of the population. In the absence of any comprehensive survey of national morbidity a useful guide is provided by an analysis of the reasons for absence from work and visits to the general practitioner since the diseases which cause major dislocation of human effort, the loss of efficiency and of working hours are not necessarily killing diseases.

Fig 69[7] shows the annual average number of days of incapacity for men aged 16 to 44 years. Scotland and north-east England have the worst record with 14½ to 16½ days' absence from work each

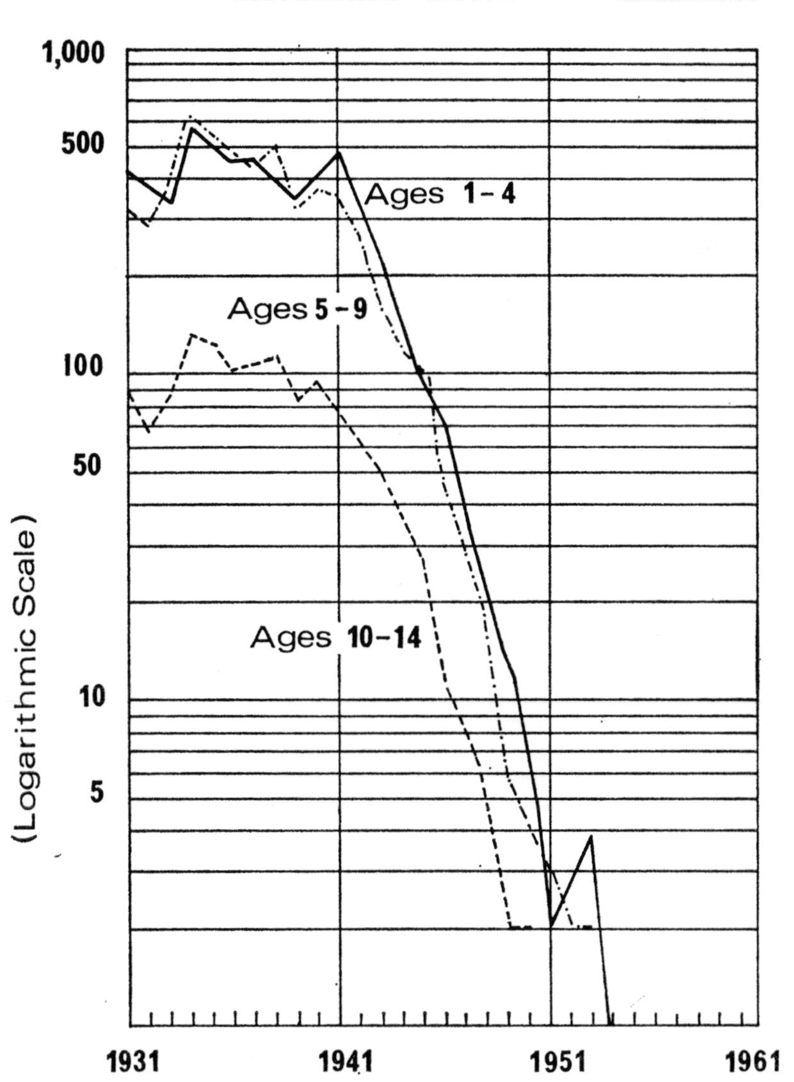

Fig 65 Diphtheria: child death rates per million in England and Wales, 1931–60 (*after Office of Health Economics 1962*)

year; eastern and south-eastern England (excluding London) have the best with 5 to 6½ days' absence. The main reasons given for sick leave are accidents, digestive difficulties, psychoses, arthritis and rheumatism, neurosis, nose and throat infections, tuberculosis, influenza and bronchitis. For the 45 to 64 years age group the distribution pattern and reasons for absence are somewhat different (Fig 70). Scotland and north-eastern England have the highest average number of days' absence (20½ to 22½ days) and eastern and southern England the lowest (10 to 12½ days). The reasons include bronchitis, influenza, arthritis and rheumatism,

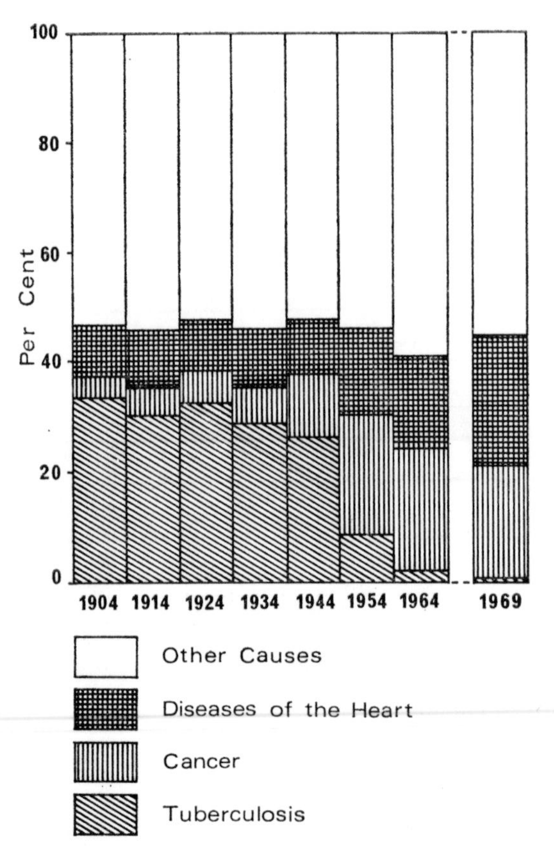

Fig. 66 Deaths in Britain by cause, 1904–69 (*Registrar-General*)

heart disease, digestive complaints, accidents, anxiety and other neuroses, tuberculosis, 'stroke' and high blood pressure.[8]

The pattern for females aged 16 to 44 years is different from that for males in the corresponding age group (Fig 71). Wales has the highest average number of days' absence from work per annum ($12\frac{1}{2}$ to $14\frac{1}{2}$ days); south-west, southern, and eastern England, the North Midlands and the East and West Ridings of Yorkshire are the regions where the absenteeism is least ($6\frac{1}{2}$ to $8\frac{1}{2}$ days). Reasons for their sick leave include psycho-neurotic disorders, nose and throat infections, complications of preg-

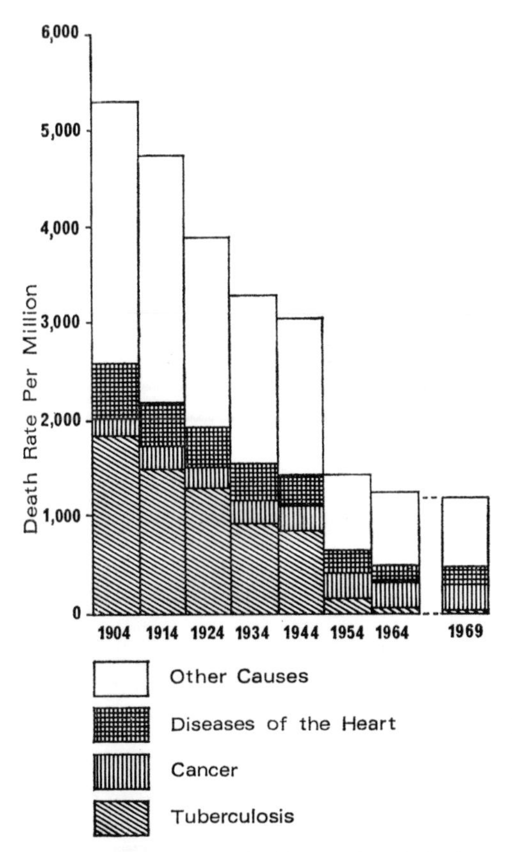

Fig 67 Death rates in England and Wales, 1904–69 (*Registrar-General*)

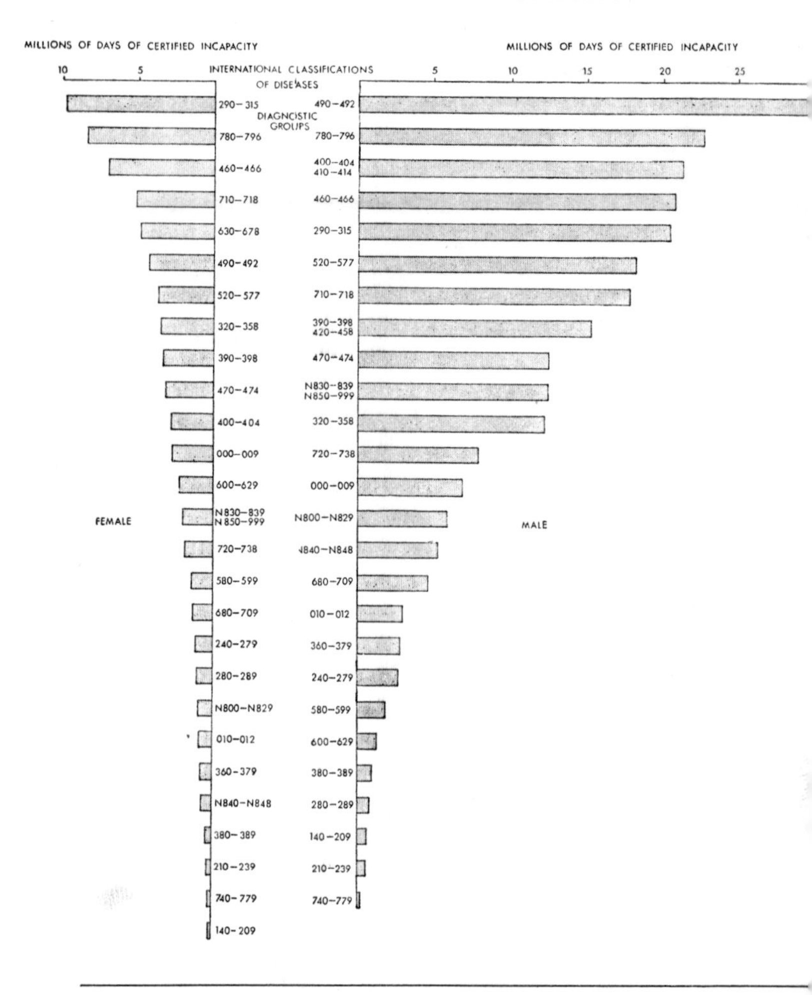

nancy,[9] arthritis and rheumatism, psychosis, bronchitis, accidents, tuberculosis, and influenza. Women belonging to the 45 to 64 age group in Lancashire, Westmorland, and Cumberland lose more time from work (15 to 16 days a year) than elsewhere in the country; those in south-western, southern, and south-eastern England and the North Midlands absent themselves from work on average only 9½ to 12 days a year (Fig 72). The main reasons for the sick leave are arthritis and rheumatism, psycho-neurotic

I.C.D. number	Diagnostic groups according to the *International Classification of Diseases, Injuries and Causes of Death* (8th Revision)
010-012	Tuberculosis of respiratory system
000-009	Other infective and parasitic diseases
013-136	
140-209	Malignant neoplasms including neoplasms of lymphatic and haematopoietic tissue
210-239	Benign neoplasms and neoplasms of unspecified nature
240-279	Endocrine, nutritional and metabolic diseases
280-289	Diseases of blood and blood-forming organs
290-315	Mental disorders
320-358	Disorders of nervous system
360-379	Diseases of the eye
380-389	Diseases of the ear and mastoid process
400-404	Hypertensive and ischaemic heart disease
410-414	
390-398	Other diseases of circulatory system
420-458	
470-474	Influenza
490-492	Bronchitis, emphysema
460-466	Other diseases of respiratory system
480-486	
493	
500-519	
520-577	Diseases of digestive system
580-599	Diseases of urinary system
600-629	Diseases of breast and genital system
630-678	Complications of pregnancy, childbirth and the puerperium
680-709	Diseases of the skin and subcutaneous tissue
710-718	Arthritis and rheumatism except rheumatic fever
720-738	Other diseases of the muscoloskeletal and connective tissue
740-779	Congenital anomalies and certain causes of perinatal morbidity and mortality
780-796	Symptoms and ill-defined conditions
N800-N829	Fractures
N840-N848	Sprains and strains of joints and adjacent muscles
N830-N839	Other accidents, poisonings and violence
N850-N999	

Fig 68 Working days lost in Britain per 1,000 insured population (of equivalent 1951 age distribution) by selected causes, 1960–1

disorders, high blood pressure, bronchitis, heart disease, accidents, gynaecological disorders and digestive complaints. It is an interesting feature that married women appear not to lose more time from work than men despite family commitments. For the older men and women absences are generally attributable to heart disease, crippling rheumatism, and neurosis. Infectious diseases are no longer important as a reason for absence from work.

A further guide to certain aspects of social malaise in Britain is the number of visits made by patients to their doctor. Fig 73 shows the annual average number of visits to the doctor made by males 16 to 44 years old. Wales has the highest number (5 to 6 visits); southern and eastern England the lowest number (2 to 3 visits). The commonest reasons for seeing the doctor were treatment for accidents, nose and throat infections, skin troubles, digestive difficulties, influenza, anxiety and neurosis. In the case of females 16 to 44 years (Fig 74), Wales, Lancashire, Westmorland, and Cumberland record the highest number of visits to the doctor (6 to 7). South-western and southern England and the North Midlands the lowest (3 to 4). The reasons for the visits were mainly during pregnancy and for nose and throat infections, anxiety and neurosis, gynaecological disorders, accidents and skin troubles. During middle age both men and women seek treatment rather more frequently. Males, 45 to 64 years, in Scotland, north-east England and Wales visit the doctor 7 to 8 times a year; In Dorset, Hampshire, Berkshire, and Oxfordshire the average number of visits is less than 3 (Fig 75). These visits are primarily for bronchitis, heart disease, accidents, digestive difficulties, arthritis and rheumatism, and nose and throat infections. For middle-aged women 45 to 64 years (Fig 76) the numbers of visits are highest in Scotland, Wales, northern and north-western England (8 to 9 visits) and lowest in Oxfordshire, Berkshire, Hampshire, and Dorset (4 to 5 visits). The commonest reasons for seeing the family doctor are arthritis and rheumatism, anxiety and neurosis, nose and throat infections, bronchitis, gynaecological disorders, accidents and blood pressures.

During the last two decades there has been a noticeable downward trend in respiratory tuberculosis, rheumatism, appendicitis, peptic ulcer, asthma, pleurisy, anaemias and diseases of the skin as causes of sickness absence from work. This same period has witnessed an increase in mental illness and accidents. Diabetes and arteriosclerosis have shown an increase among men.

Notes to this chapter are on pp 258–9.

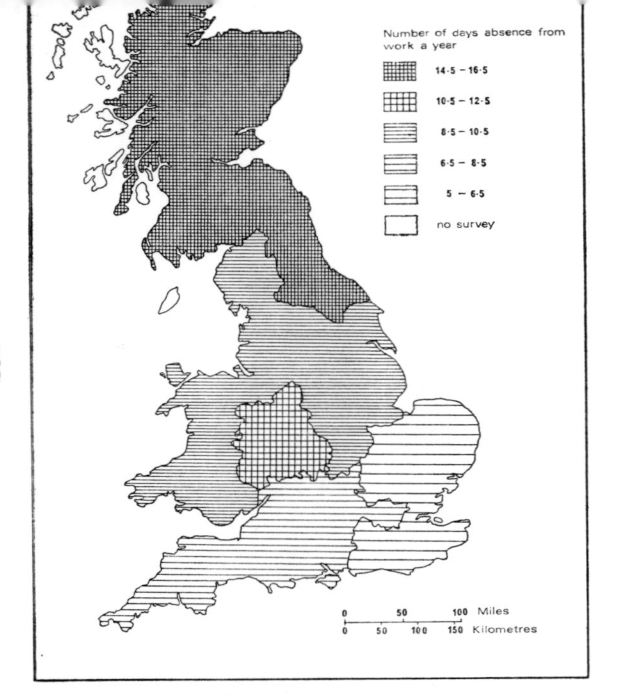

Fig 69 Annual average number of days' absence from work through sickness, males, aged 16 to 44 years (*adapted from* Reader's Digest *Atlas of the British Isles*)

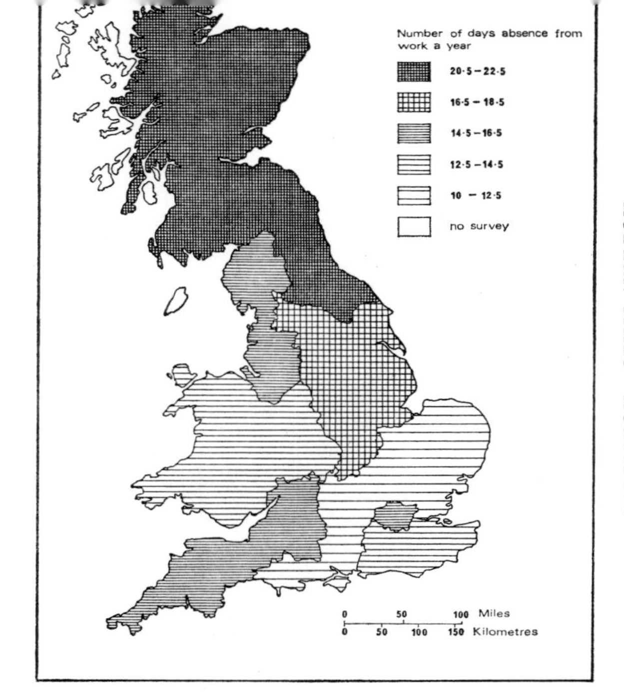

Fig 70 Annual average number of days' absence from work through sickness, males, aged 45 to 64 years (*adapted from* Reader's Digest *Atlas of the British Isles*)

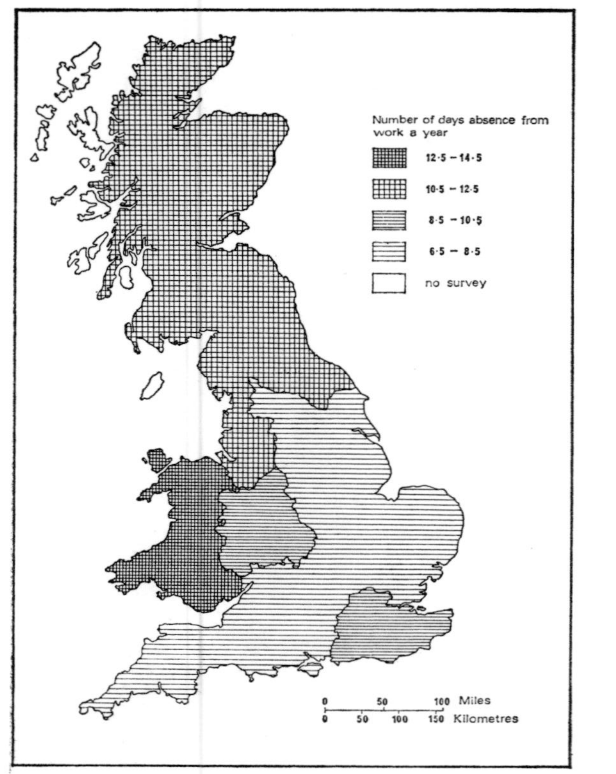

Fig 71 Annual average number of days' absence

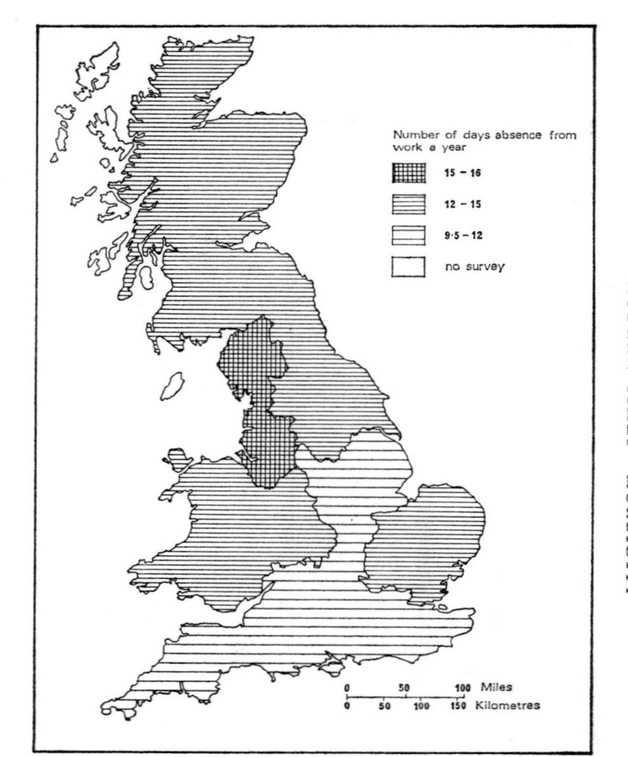

Fig 72 Annual average number of days' absence

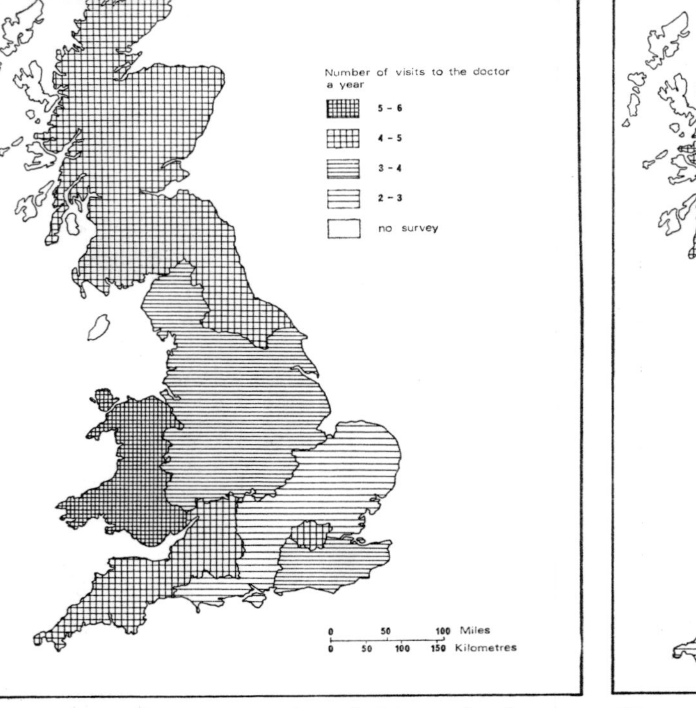

Fig 73 Annual average number of visits to the doctor, males, aged 16 to 44 years (*adapted from Reader's Digest Atlas of the British Isles*)

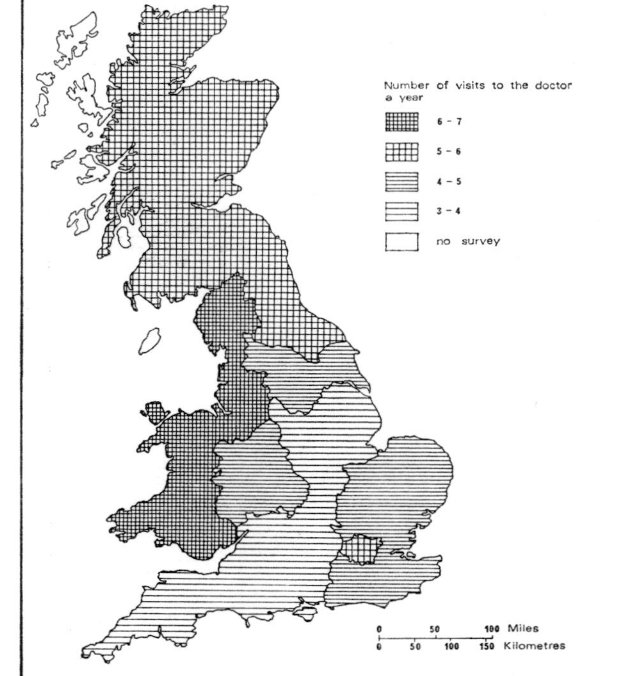

Fig 74 Annual average number of visits to the doctor, females, aged 16 to 44 years (*adapted from Reader's Digest Atlas of the British Isles*)

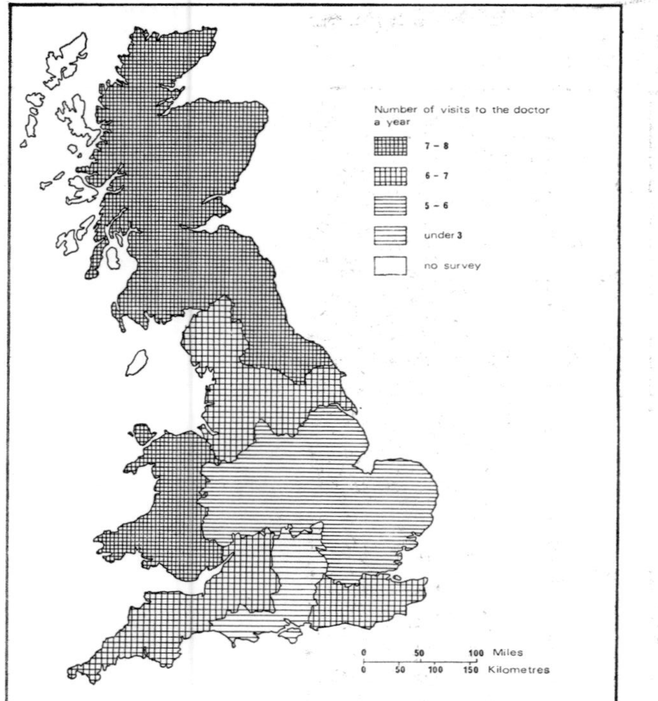

Fig 75 Annual average number of visits to the doctor, males, aged 45 to 64 years (*adapted from*

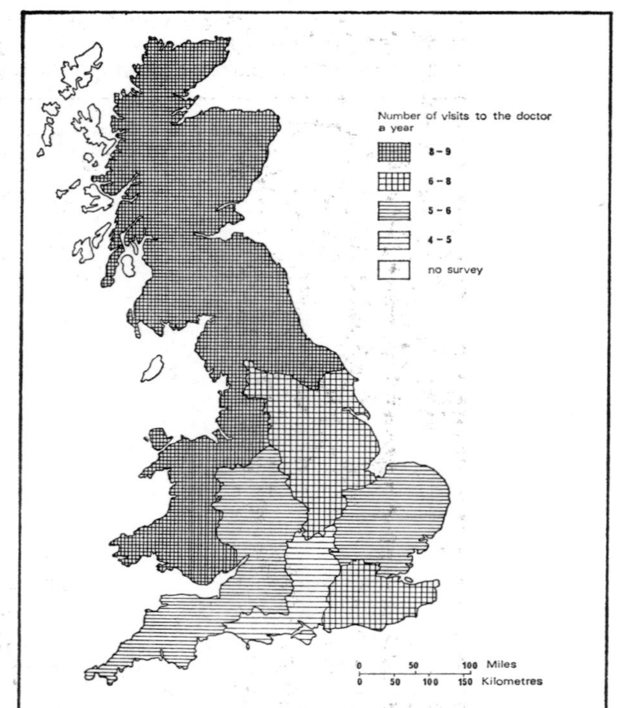

Fig 76 Annual average number of visits to the doctor, females, aged 45 to 64 years (*adapted from*

Plate 15 Group of Glasgow children with rickets *c* 1910

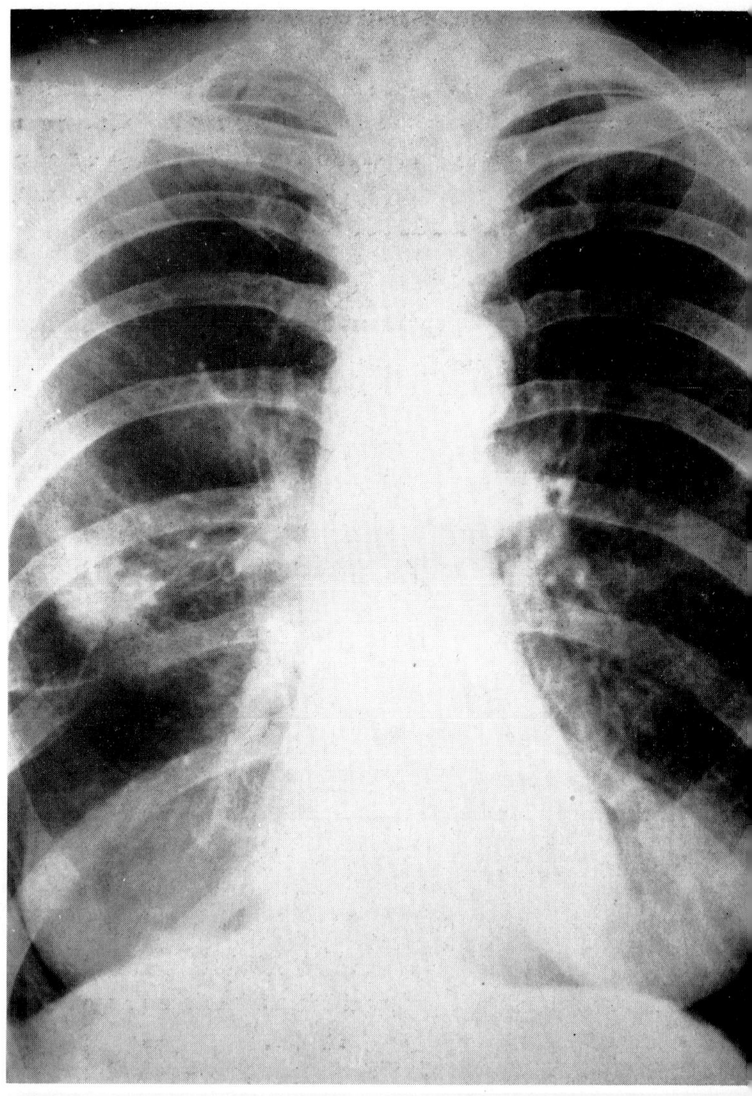

Plate 16 Lung cancer X-ray photograph of chest showing a carcinoma of the upper part of the left lung

14

Modern Times—Mortality

Mortality indices provide a measure of incidence for those diseases which have a high fatality rate. For those in which the fatality rate is low, as for example, 'rheumatoid arthritis', they are of necessity almost meaningless. Fig 77 shows the average annual deaths in the United Kingdom from the main causes, for both males and females, for the period 1959–63. The major causes of death for men are ischaemic heart disease, followed by vascular lesions affecting the central nervous system. For females, vascular lesions affecting the central nervous system is the major cause,

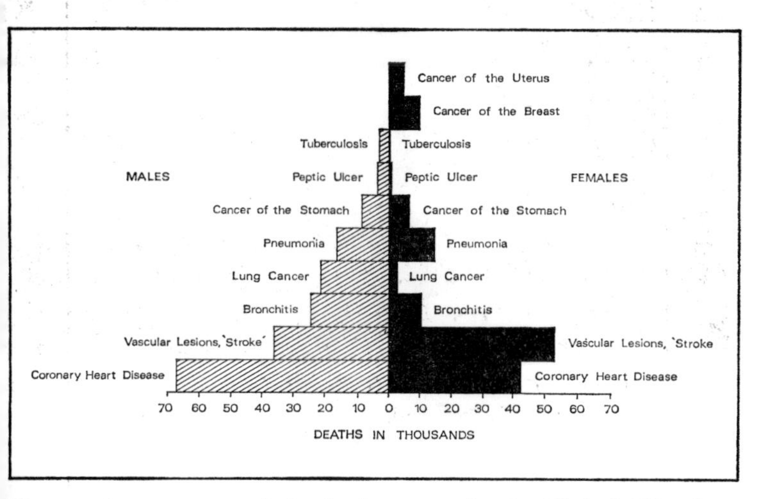

Fig 77 Average annual deaths by cause in the United Kingdom, 1959–63

followed by coronary heart disease. Bronchitis and lung cancer are much more severe for men than women. The substantial number of female deaths from cancer of the breast and cancer of the uterus redress the balance between the sexes. Deaths from pneumonia and stomach cancer are roughly the same in both sexes.

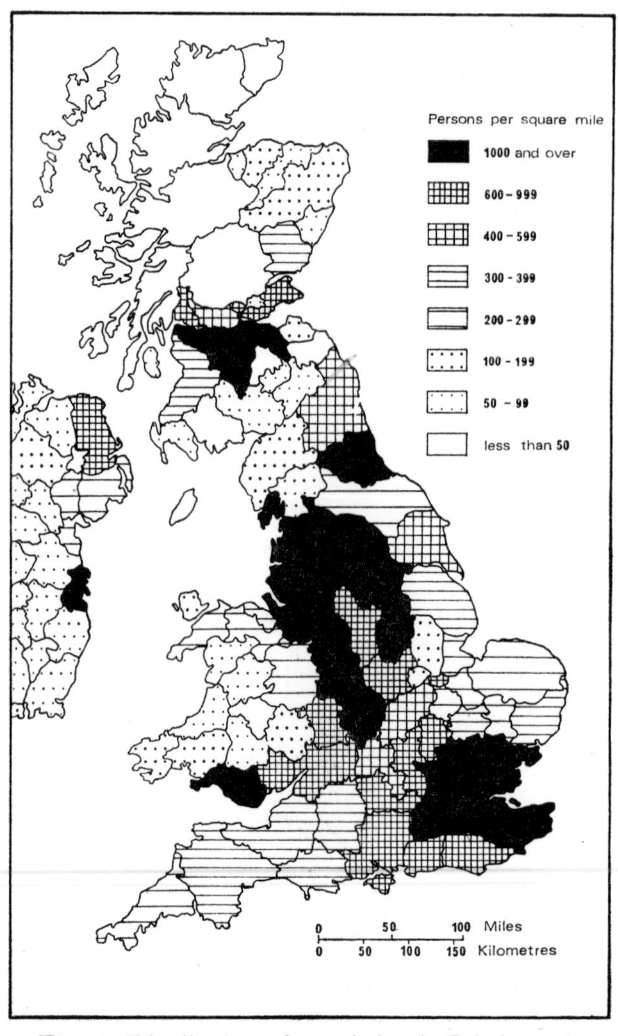

Fig 78 Distribution of population in Britain, 1961

The distribution of population in Britain in 1961 is given in Fig 78 and the regional incidence of mortality from a selection of major causes, for *men* only, for the period 1959–63 inclusive is given in Figs 80, 82–87, 89 and 91.[1] Demographic[2] base maps (which relate mortality to actual numbers of people 'at risk' to disease in local areas) rather than geographical base maps are employed to present the mortality data (Fig 79). The 'squares' on the maps represent metropolitan boroughs, county boroughs (Scotland: 'cities of counties') and aggregates of municipal boroughs and urban districts (Scotland: large and small burghs) of the various administrative counties; the 'diamonds' represent the aggregates of rural districts (Scotland: 'landward'). In each case the areas of the 'squares' and 'diamonds' are proportional to the populations they represent (Fig 79). Standardised mortality ratios (SMR) rather than crude death rates are used on the maps because the ratios make allowance for the fact that some areas have more than the average number of older people in their populations while other areas have comparatively youthful populations.[3]

Ischaemic heart disease

Annual deaths from ischaemic heart disease in the United Kingdom during the period 1959–63 averaged 66,500 males and 42,000 females. A line drawn from the Severn Estuary to the estuary of the Humber divides the country into two contrasting regions (Fig 80). The southern half has low mortality ratios for ischaemic heart disease, the northern half has high ratios for the same cause. There are major areas of particularly unfavourable mortality experience in Scotland, Northern Ireland, Tyneside, Teesside, Lancashire, the West Riding of Yorkshire, and South Wales.

Internationally, ischaemic heart disease correlates with urbanised, industrialised societies, the incidence of the disease being more frequent in regions of affluence. Within Britain, the problem is seemingly the reverse. The mortality ratios are lower in the economically favoured south and east of England (Fig 81). Some epidemiologists suspect social promotion as a risk factor for ischaemic heart disease and would suggest that the lower ratios of

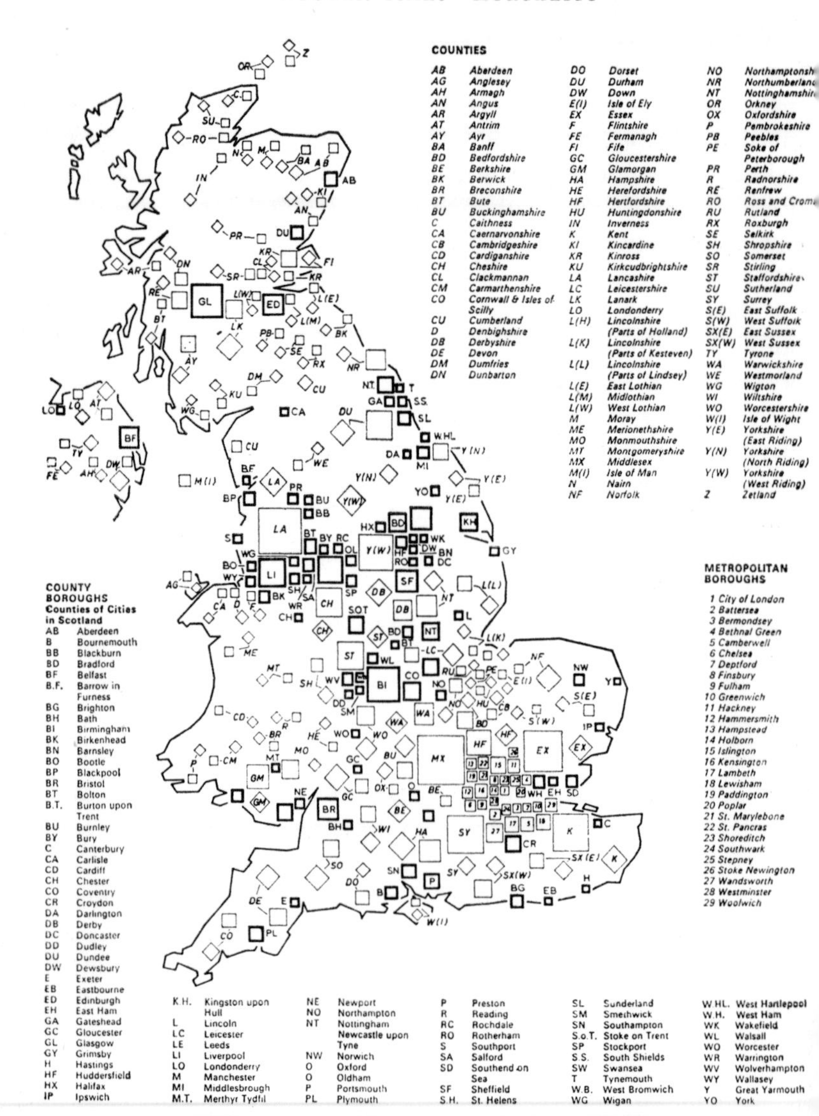

Fig 79 Demographic location map of the United Kingdom. The areas of the squares and diamonds are proportional to the populations they represent

the south-east of the country reflect adaptation to longer-lasting prosperity. Others note a high incidence of heart disease in men of strong drive and energy, imbibed with a competitive spirit and sense of urgency. It seems unlikely, however, that one part of the

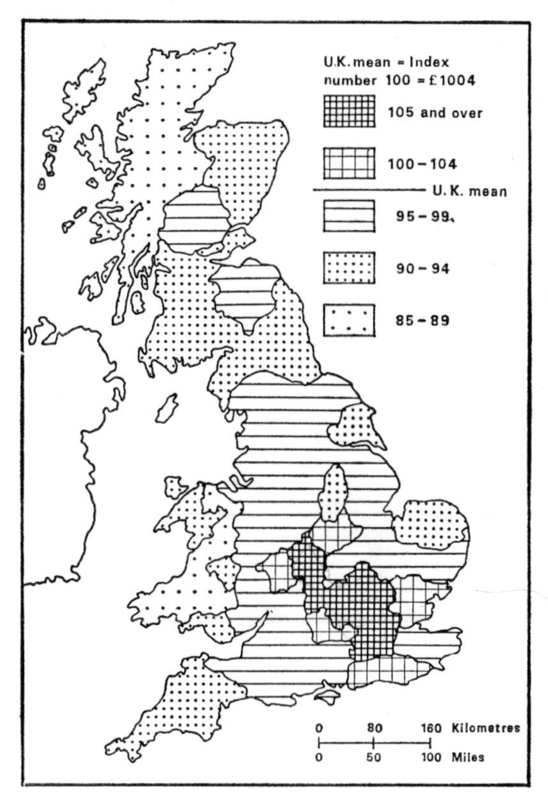

Fig 80 Affluence, as indicated by total net income, 1964–5 (*after Rawstron and Coates 1971*)

country has a monopoly of men of this behavioural type. Other suspected causes of ischaemic heart disease include diet (especially a high intake of saturated fats), cigarette smoking, high blood pressure, and lack of physical activity. Diets and eating patterns in the different parts of Britain vary widely (*see* Fig 24). In the

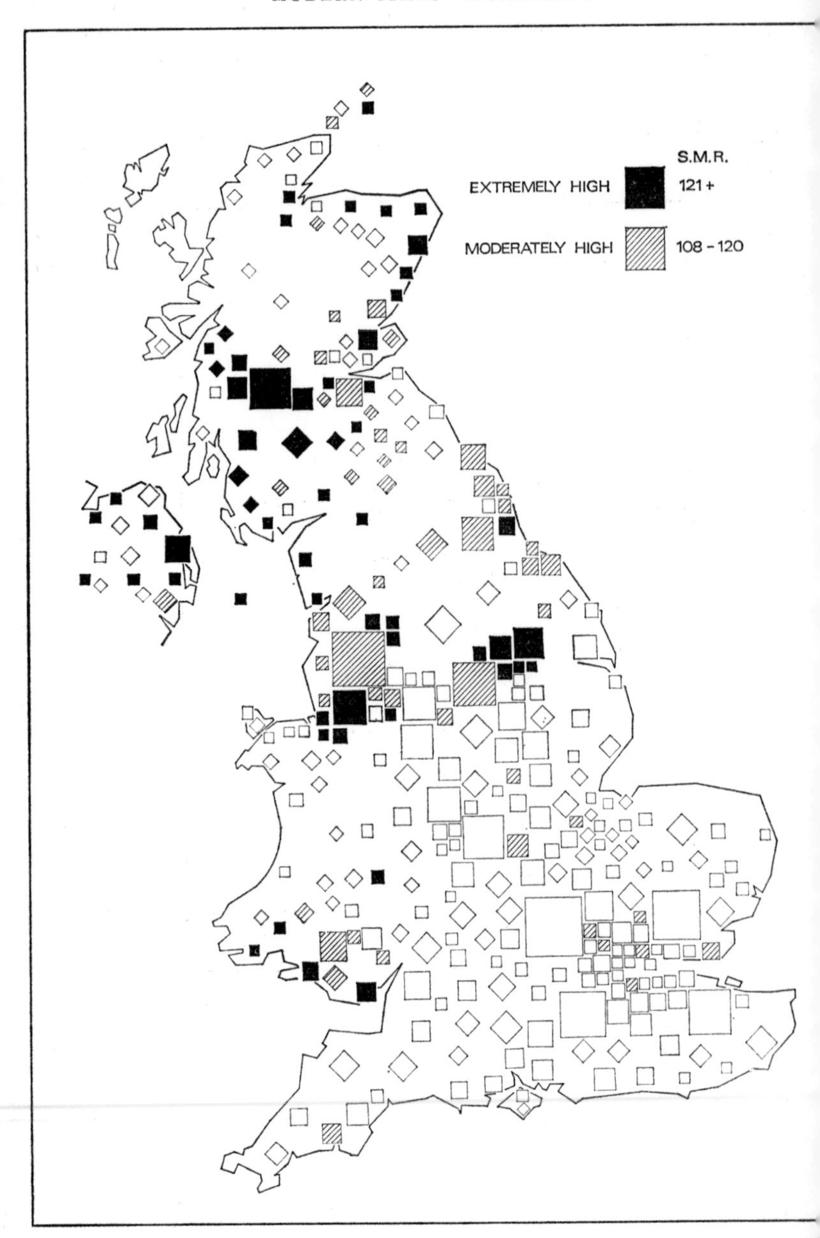

Fig 81 Ischaemic heart disease: distribution of mortality in the United Kingdom, males, 1959–63 (*simplified from Howe 1970*)

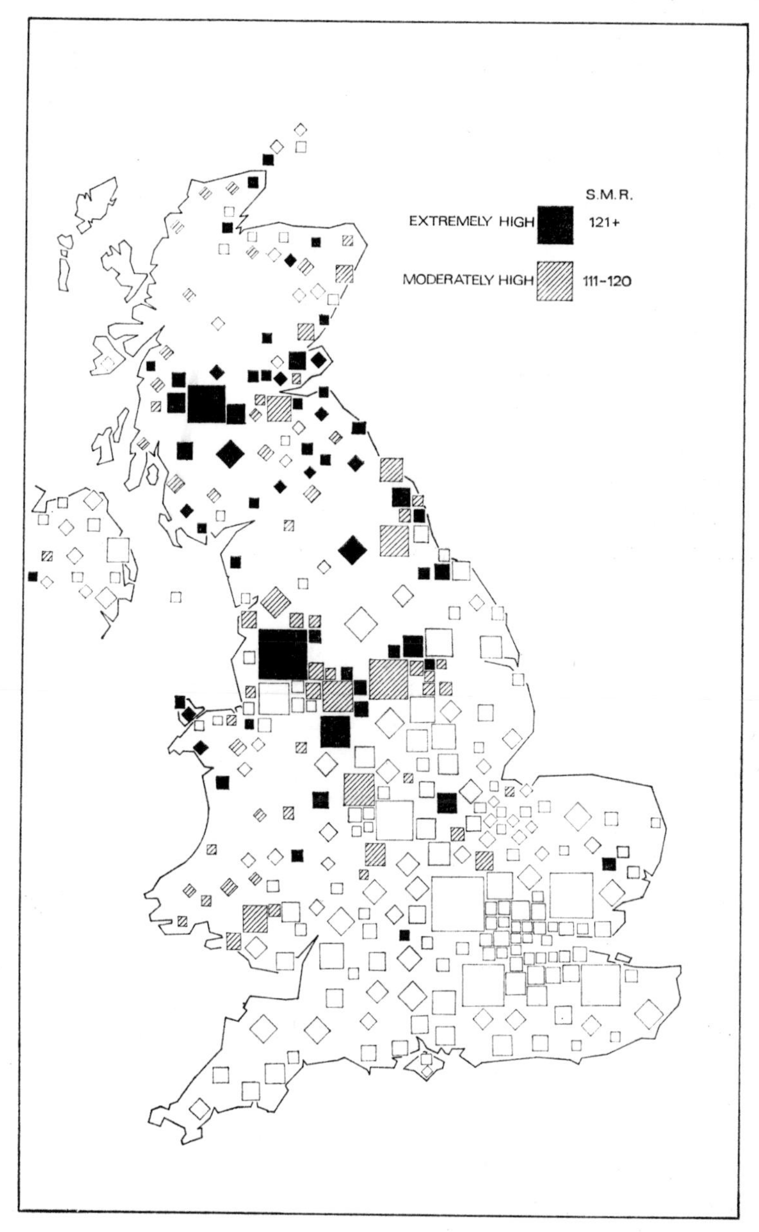

Fig 82 Cerebro-vascular disease: distribution of mortality in the United Kingdom, males, 1959–63 (*simplified from Howe 1970*)

London area household consumption is well above the national average for all selected items of food except margarine, cakes, and biscuits. In Scotland on the other hand the consumption of cakes, biscuits, margarine, beef, and veal is in excess of the national average and well below for all other selected items of food. In the Midlands the diet is characterised by an excess of pork, mutton, and lamb, in Wales by an excess of butter. In this context associations between dietary fats, blood cholesterol, atherosclerosis and ischaemic heart disease, and between high intake of sugar and ischaemic heart disease are postulated. The disease is also thought to be inversely correlated with hardness of the water supply (Fig 20), yet a detailed study within the city of Glasgow, every ward of which receives a soft water supply from Loch Katrine, reveals mortality experience in certain of the wards 20 to 30 per cent below the national average.[4] Doubtless several interrelated factors are implicated in ischaemic heart disease (*see* p 5).

Cerebro-vascular disease

Practically 36,400 men and 52,900 women died from 'stroke' each year in the United Kingdom during the period 1959–63. The distributional pattern of mortality (Fig 82) is somewhat similar to that for ischaemic heart disease. This is not altogether surprising since high blood pressure and atherosclerosis both play a part in each condition. Mortality ratios are above the national average in the north and west of the country and below the national average in the south and east. Some of the regional variations in 'stroke' ('shock') mortality may be a reflection of variations in clinical diagnosis or the availability and use of treatment for high blood pressure. The presence or absence of specific trace elements or 'micro-nutrients' may be responsible for vascular damage (p 32). Some trace elements are ubiquitous, others have an uneven distribution.

Chronic bronchitis

About 24,000 men and 10,000 women die each year from chronic bronchitis in the United Kingdom. Areas of most unfavourable experience from this disease include parts of London, south-west

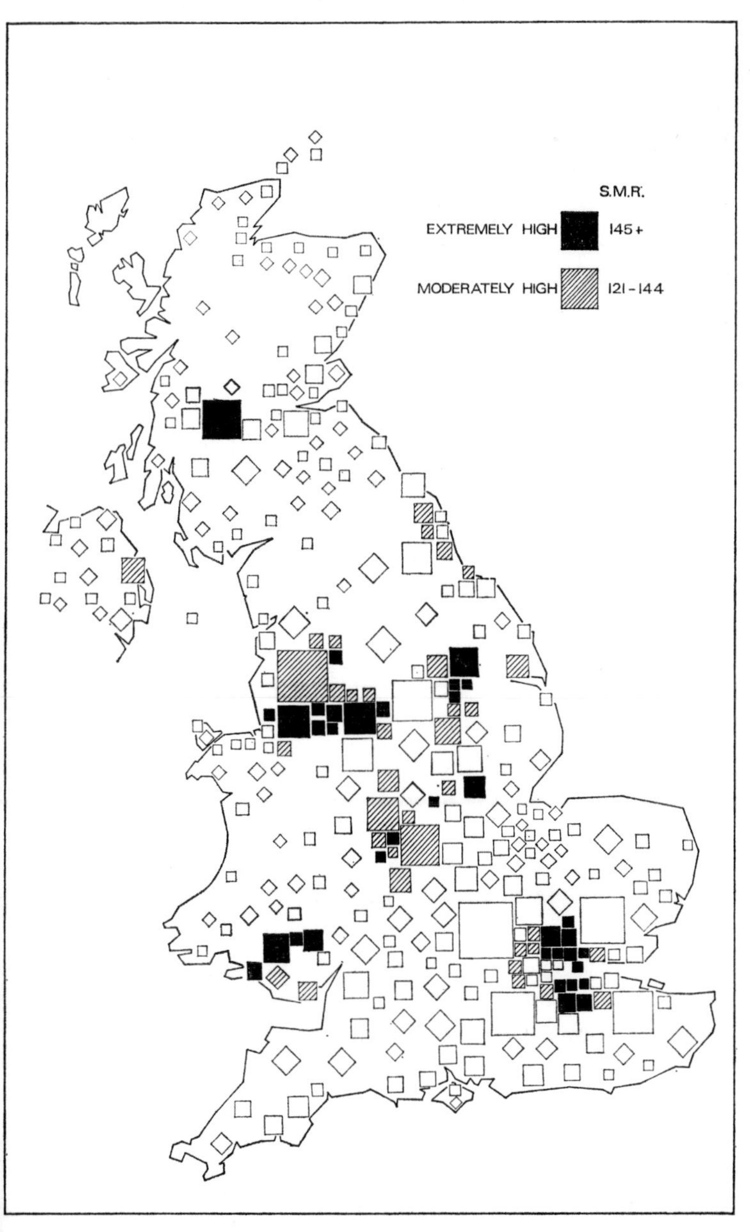

Fig 83 Chronic bronchitis: distribution of mortality in the United Kingdom, males, 1959–63 (*simplified from Howe 1970*)

Lancashire and Merseyside, the Midlands, the West Riding of Yorkshire, South Wales and West Central Scotland (Fig 83). The origins of the disease are not fully understood but the evidence of the distributional pattern suggests that it is a town disease. It appears to correlate with heavily industrialised and polluted areas; rural areas with slight air pollution seem to escape. In the industrialised areas the atmosphere is polluted by smoke from domestic and factory chimneys, gases from a wide variety of factories, motor vehicles, railway locomotives, etc (p 58 and Figs 26 and 27). Nevertheless it has also been observed that unskilled workers and their wives suffer the highest mortality, the rate falling progressively to the lowest among professional men and their wives. Does the higher bronchitis mortality of certain of the towns depend on urban atmospheric pollution or on urban class structure and social conditions? Cigarette smoking (p 68 ff), a habit indulged in universally by both males and females throughout the country, has long been considered to be a contributory causal factor in chronic bronchitis. Some constitutional or genetic element in susceptibility may also be implicated. Chronic bronchitis itself is not a bacterial disease although it is normally associated with intermittent bacterial infection.

The worst episode of bronchitis mortality so far recorded in Britain was during the great London smog of December 1952, when concentration of sulphur dioxide reached 134 parts per 100 million (p 59). In four days there were nearly 4,000 deaths, largely attributed to bronchitis. Deaths occurred mainly among elderly persons suffering from chest or heart disease. Such mortality was as dramatic as the worse days of cholera in the nineteenth century (p 168) or of plague in the fourteenth and seventeenth centuries (p 93 ff and p 120 ff).

Lung-bronchus cancer

Male deaths from lung-bronchus cancer in the United Kingdom during the period 1959–63 amounted to 21,900 each year. For women the figure was 3,700. There were practically six male deaths for every female death from this disease. The distributional pattern shows four major areas of particularly unfavourable

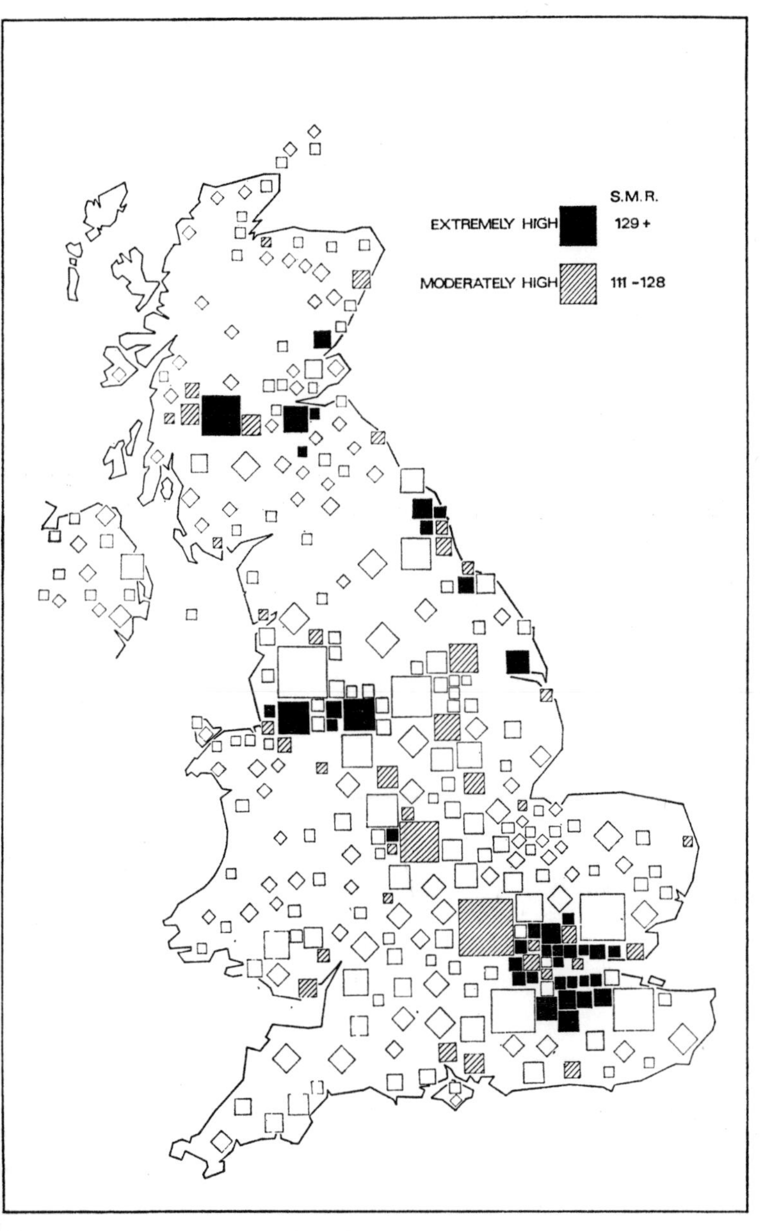

Fig 84 Lung-bronchus cancer: distribution of mortality in the United Kingdom, males, 1959–63 (*simplified from Howe 1970*)

mortality, London, Merseyside–south-east Lancashire, north-east England and Central Scotland. There are two lesser concentrations of high mortality in West Bromwich and in Kingston-upon-Hull (Fig 84).

A marked urban–rural gradient in mortality ratios is evident and in the main people living in areas of heavy industry are most severely affected. Smoking habits (p 68 ff), industrial carcinogens, atmospheric pollution (p 57 ff) and diagnostic facilities are among the multiplicity of contributory factors.

Data is not available to indicate any regional variation in the intensity of cigarette smoking although it is known[5] that the consumption of tobacco products in Scotland is higher in both sexes than in England and Wales, whether considered 'per adult' or 'per smoker'. The figures for consumption of manufactured cigarettes per week show substantial differences.

Table 10 Mean consumption of manufactured cigarettes per week in England–Wales and in Scotland

	Per adult		*Per smoker*	
Country	*Male*	*Female*	*Male*	*Female*
England and Wales	68	36	128	84
Scotland	80	37	142	97

On a yearly basis the Scottish male cigarette-smoker consumes an average of about 7,400 cigarettes, the English and Welsh about 6,700. The corresponding figures for women would be about 5,000 and 4,400. The difference between Scotland on the one hand and England and Wales on the other, is about 11 per cent for men and 15 per cent for women. Smokers in Scotland not only consume far more cigarettes per head than smokers in England and Wales but the proportion of non-tipped cigarettes smoked is also higher (plate p 222).

Stomach cancer

Practically 8,800 men and about 6,900 women die each year from stomach cancer. Lancashire–Cheshire, north-east England, South

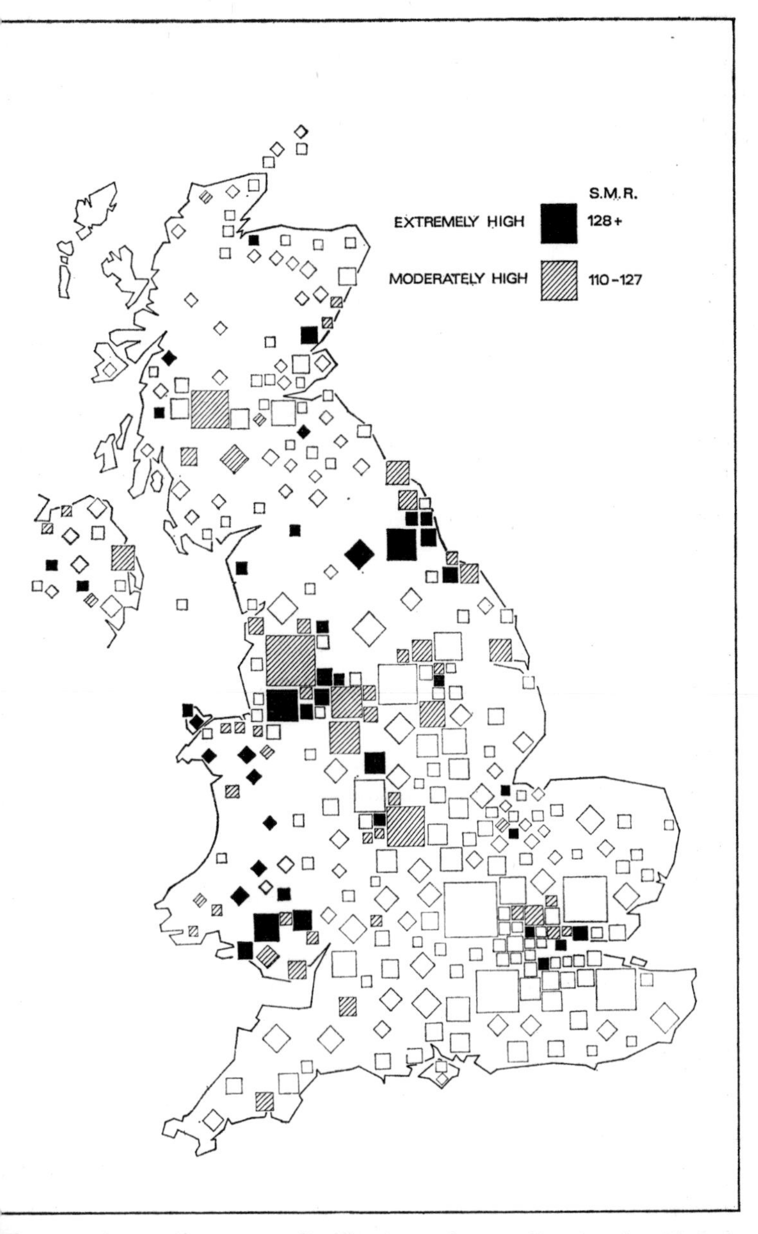

Fig 85 Stomach cancer: distribution of mortality in the United Kingdom, males, 1959–63 (*simplified from Howe 1970*)

Wales and the Potteries are major areas of unfavourable mortality experience from this disease. Extremely high ratios also occur sporadically in parts of the Midlands, the London area, Central and North Wales, Scotland, Northern Ireland, and the Fenlands, but populations in these areas are relatively small (Fig 85).

No particular urban or rural association is suggested by the distributional patterns. The relationship, be it direct or indirect, is probably with a factor or factors common to both the urban and rural environment, or else is genetic. The numbers of factors which have been postulated as being associated with stomach cancer in different parts of the world is very great. The disease has been associated with, for example, alcohol, with smoked or charcoal-cooked foods, with diets in which cereals form a large part and fats, fruit and vegetables are rarely eaten (this may merely indicate low socio-economic status), with the absence of vital trace elements in the soil, with chemically polluted drinking water and with hereditary predisposition. Gastric carcinoma is thought to be significantly higher in people with blood group A and non-secretors although this does not appear to be particularly evident when Figs 4 and 85 are compared.

Pneumonia

Annual deaths from broncho- and lobar-pneumonia during the 1959–63 period averaged 16,180 men and 17,200 women. The distributional pattern is dominated by five areas of extremely high ratios—London, Lancashire, Yorkshire, Durham, and the Midlands (Fig 86).

It would be expected that the disease would take its highest toll generally in those densely-populated, urban areas where the stress of life is high and the standard of living accommodation is low. It seems likely, however, that other factors may operate including socio-economic standards and the presence of influenzal or other predisposing infections.

The successful treatment of lobar-pneumonia is one of the great triumphs of modern medicine. Broncho-pneumonia, with its many different bacteria is not so readily amenable to treatment with any single anti-bacterial agent.

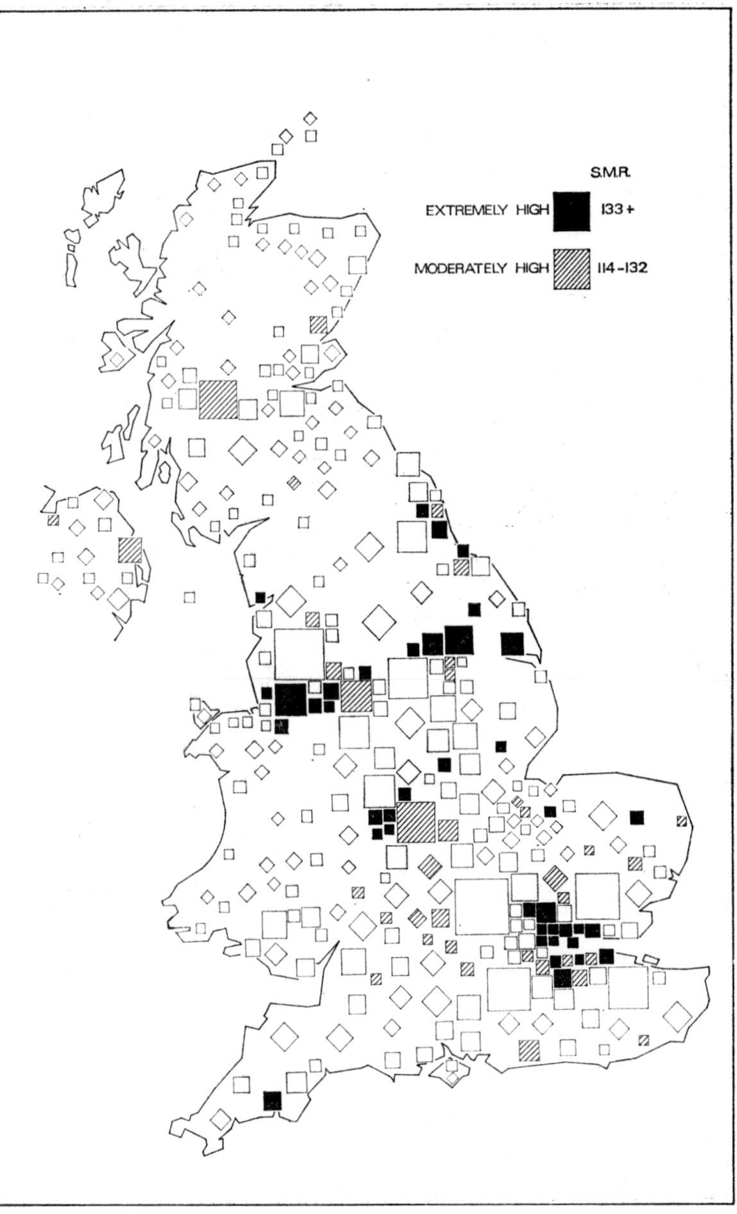

Fig 86 Pneumonia: distribution of mortality in the United Kingdom, males, 1959–63 (*simplified from Howe 1970*)

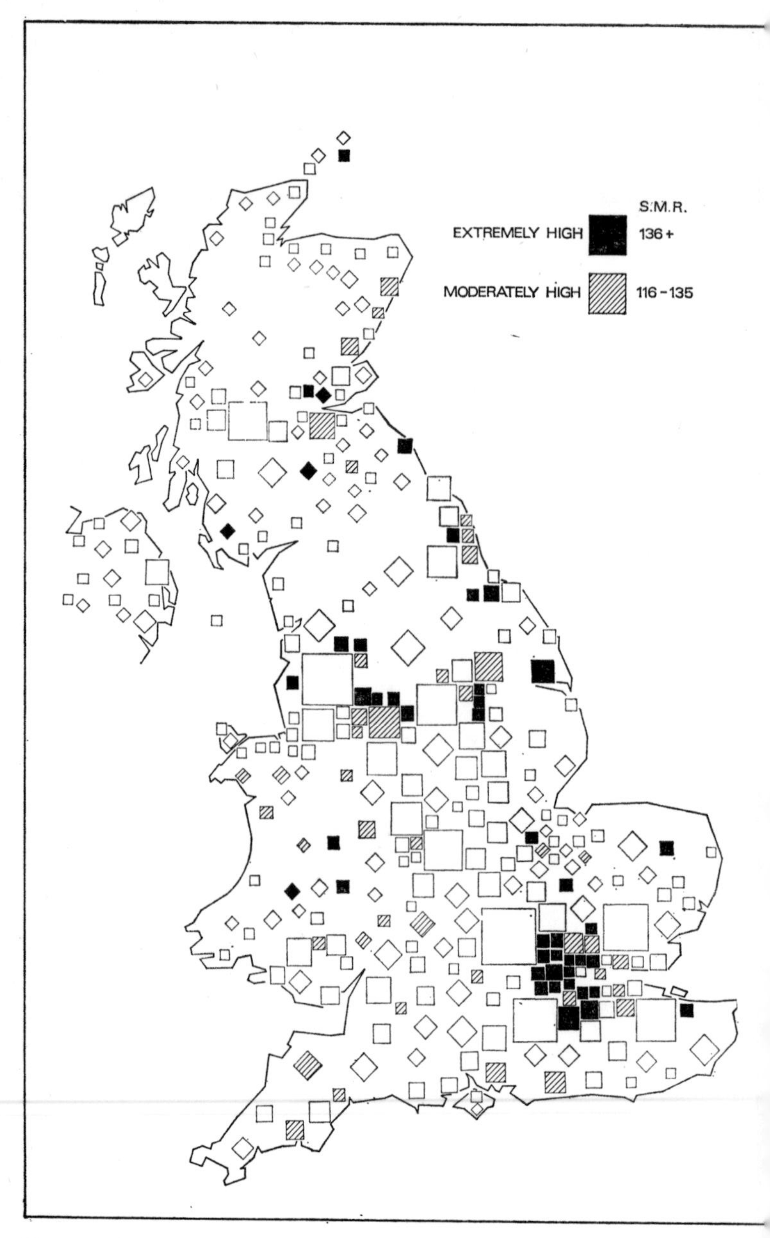

Fig 87 Suicide: distribution of mortality in the United Kingdom, males, 1959–63 (*simplified from Howe 1970*)

Suicide

About 3,470 men and 2,400 women deliberately killed themselves every year during the period 1959–63. There is a pronounced concentration of suicides in London, South Lancashire, several parts of Yorkshire and in Teesside–Tyneside. There are other areas, however, which include Cambridge and Canterbury (Fig 87).

The aetiology of suicide is most complex but the importance of social and cultural factors is well established. There are a number of vulnerable groups of people in which the suicide rate is well above average. These include psychotics, persons suffering from manic depressive illness, those who are temperamentally inadequate for the stresses and strains of life, alcoholics, and drug addicts. Chronic ill-health, old age and the dread of becoming helpless, and loneliness are other factors. In cities the loneliness of individuals is regrettably exacerbated by the proximity of large numbers of people. Death rates per million living by age groups and by sex, are given in Fig 88.

Infant mortality (both sexes)

Annual deaths of children (both sexes) under one year old in the United Kingdom during the period 1959–63 averaged 21,055. The infant mortality rate was 22·5 per 1,000 live births but there was considerable regional variation. The overall pattern of infant mortality rates is clearly one of high rates in the north and west of the country and low rates in the south and east (Fig 89) but with appreciable variation above and below the national average even at the local level.

In general, infant mortality is a disease of poverty and the areas with high rates are characterised by high levels of unemployment, low wages, large families, and low educational standards. Overcrowding, poor hygiene, and faulty diets are the usual consequences of poverty.

All Causes

The overall reduction in All Causes death rates in Britain since about 1850 is highlighted by the decrease in mortality rates among young adults of both sexes aged 15 to 44 years. It is largely upon this group that the productivity of the country largely depends, both in economic and social terms (Fig 90). For the first ten or so years after records began in 1841 there was no reduction but there-

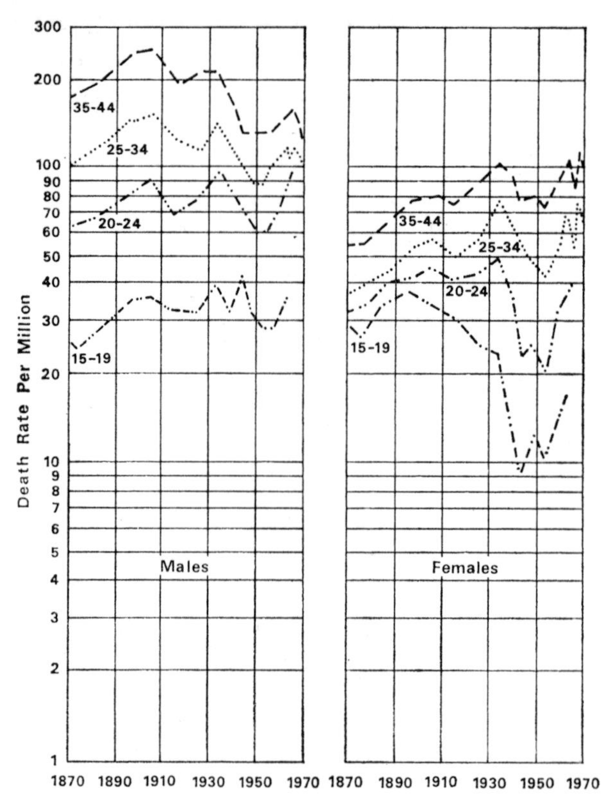

Fig 88 Suicide: death rates per million living in England and Wales, by sex and age groups, 1870–1970 (*after Office of Health Economics 1964*)

after for over a hundred years, except for the war years, there has been a continuous decrease. From the 1860s until the 1930s improving medical, economic, and social conditions and the development of the public health services brought a steady lowering of the death rates. From the 1930s to the 1950s the speed of

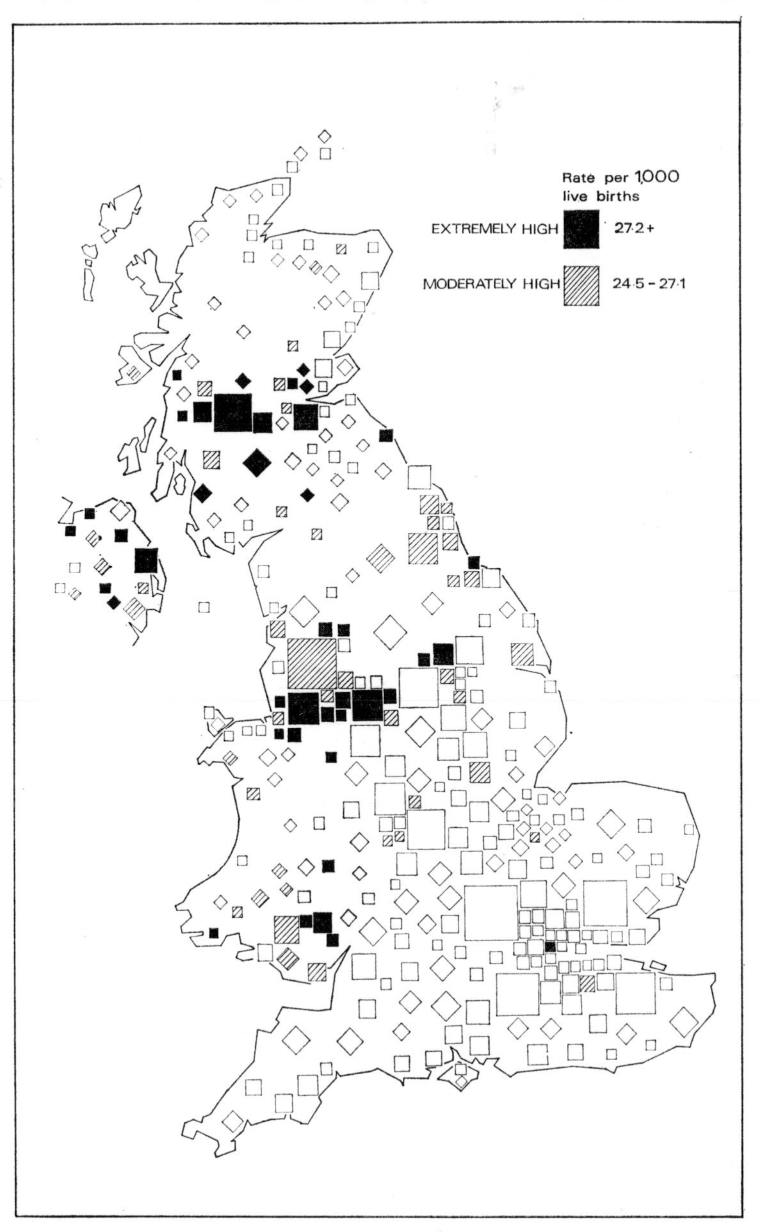

Fig 89 Infant mortality: distribution in the United Kingdom, 1959–63 (*simplified from Howe 1970*)

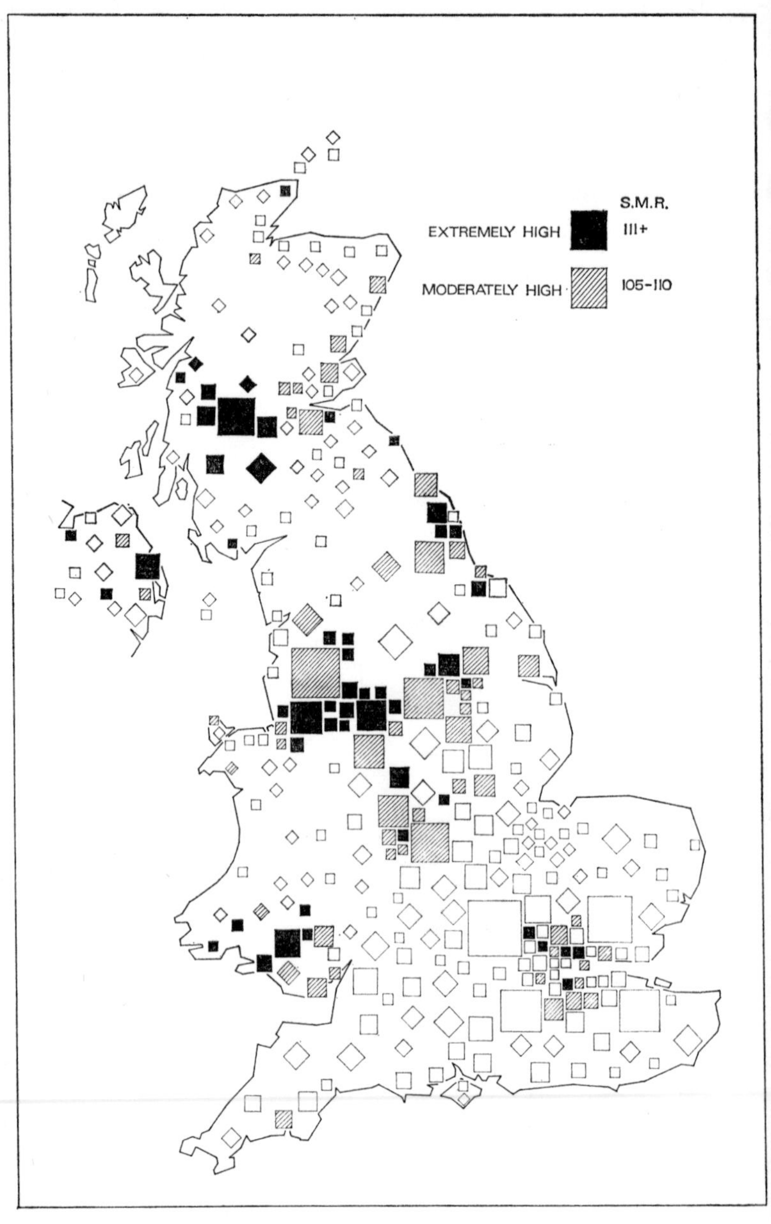

Fig 90 Mortality from 'All Causes' in the United Kingdom, males, 1959–63 (*simplified from Howe 1970*)

improvement accelerated, due largely to the control of infection by a combination of social advances, preventive medicine and chemotherapy. Since the 1950s there has been a marked falling-off in the speed of improvement and the mortality rates for 1969 showed a slight increase over those for 1968.

The summation of community responses to the total complex of environmental factors and ways of life which prevail in twentieth-century Britain and of improved medical and public health services is represented, for males only, by the standardised mortality ratios shown in Fig 91. This map shows the areas where, seemingly, environmental factors are conducive to health and long life

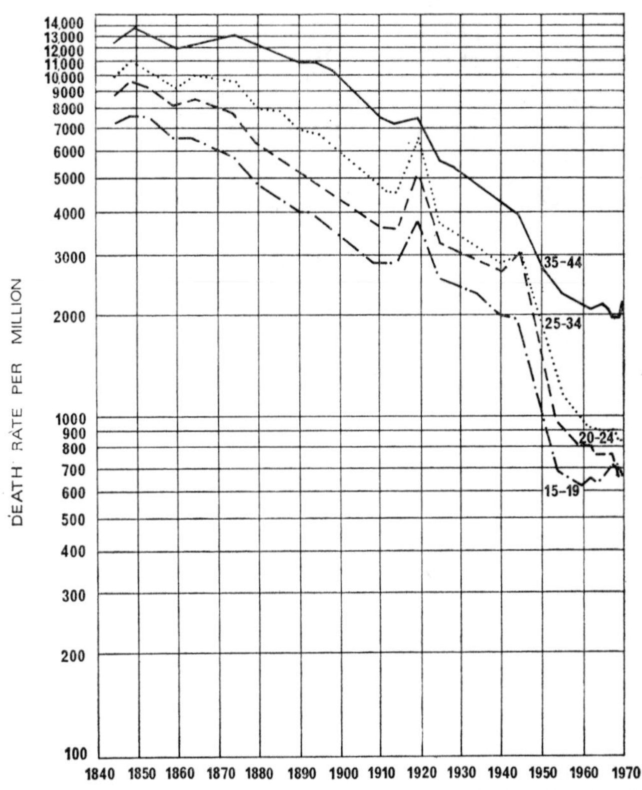

Fig 91 Death rates per million living in England and Wales, by age, 1841–1970 (*Registrar-General's Statistical Reviews of England and Wales*)

and the areas where they result in illness and death. It is a reflection of the varying degrees of human adjustment and maladjustment to environmental hazards throughout the country. Whether this geographical pattern of mortality in Britain will be maintained in the years ahead remains to be seen.

The physical environment though seemingly static is under constant interference by man. Some of the changes being wrought are imperceptible and insidious. Others are blatantly obvious. What is beyond doubt is that man is altering the balance of a relatively stable system by his actions. The ecology of life on this planet is being radically changed by him. From the health point of view some of the changes may be good, some may be harmful and others may well prove catastrophic. That there are no precise views of the impact of the changes being wrought in the balance of the great natural forces and of the new environments being created is cause for serious concern. It is palpably unwise to continue to interfere with man's habitat without, at the same time, striving to determine the real and lasting effects of such actions on man's health and general well-being. To ignore these effects may lead to the extinction of life on this planet and the death of tomorrow.

Notes to this chapter are on p 259.

Notes and References

Chapter 1, Introduction (pp 1–6)

1 World Health Organisation (WHO). Basic documents (16th ed 1965), 1
2 'An ecosystem is a functioning interacting system composed of one or more living organisms and their effective environment, both physical and biological . . .' (Fosberg, F. R., 'The island ecosystem', in Fosberg, F. R. (ed), *Man's place in the island ecosystem. A symposium* (Honolulu 1963), 1–6)
3 See Clarke, R., *We all fall down: the prospect of biological and chemical warfare* (1968) and *The science of war and peace* (1971)
4 World Health Organisation News Bulletin (June 1965). *The Practitioners*
5 For further consideration of 'Normality' see Renbourn, E. T., 'Normality in relation to human reactions', *Proc Ninth Int Congress on Industrial Medicine* (London 1948)

Chapter 2, Man in Britain (pp 7–16)

1 Offa's Dyke was an embankment and ditch built along the border hills from Prestatyn to Chepstow. It was an agreed frontier between King Offa of Mercia and the Welsh princes
2 Fleure, H. J., *A natural history of man in Britain* (Collins, London 1971)
3 See Vogel, F., 'ABO blood groups and disease', *Amer J Human Genet*, 22 (1970), 464–75 and Brothwell, D., 'Disease, micro-evolution and earlier populations: an important bridge between medical history and human biology', *Modern methods in the history of medicine*, ed E. Clarke (London 1971)
4 Weiner, A. S., 'Blood groups and disease', *Amer J Human Genet*, 22 (1970), 476–83

Chapter 3, Health Hazards of the Physical Environment (pp 17–39)

1 Winslow, C. E. A. and Herrington, L. P., *Temperature and human life* (Princeton UP, London UP, 1949)

2 Waddy, B. B., 'Climate and respiratory infections', *Lancet*, 2 (1952), 674–7

3 Boyd, J. T., 'Climate, air pollution and mortality', *Brit J Soc Prev Med*, 14(3) (1960), 123

4 Brooks, C. E. P., *The English Climate* (English Universities Press, 1954). New edition by H. H. Lamb (1964)

5 Hawkins, E., *Medical climatology of England and Wales* (Lewis, 1923)

6 See Tyler, W. F., 'Bracing and relaxing climates', *Q J Roy Met Soc*, 61 (1935), 209–15

7 Howe, G. M., 'Windchill, absolute humidity and the cold spell of Christmas, 1961', *Weather*, 17 (1962), 349–58

8 Kilogram calories per square metre per hour. This unit is used by physiologists in calculations of metabolism and body-heat output

9 See Reiter, R., 'Neuere Untersuchungen zum Problem der Wetterabhängigkeit des Menschen', *Archiv fur Meteorologie, Geophysik und Bioklimatologie*, Ser B, 4 (Vienna 1953), 327–77

10 Darby, H. C. (ed), *An historical geography of England before 1800 AD* (Cambridge 1936, 1969)

11 The trend is not readily evident in any one year or in all parts of Britain in the same year. The period 1965–8 was the wettest period of four consecutive calendar years over England and Wales since 1927–30. In each of the years 1965–8 the excess rainfall was mainly attributable to the summer six months (April–September). In 1968 south-east England experienced one and a half times its long-term average annual rainfall whereas Scotland had only 80 per cent of its average

12 m.rem is one-thousandth of a rem (rem^{10-3}). The rem is the unit of dose of damage by radiation to the human body. A dose of about 500 rem to the whole body has a 50 per cent chance of being fatal to man within a fortnight. A lifetime's accumulation in the body of natural radiation (cosmic rays, radiation from rocks, potassium 40 in the body) is about 3 rems

13 An approximate formula for gauging the dose either to bone marrow or to gonads is that the local gamma radiation dose is equal to one-half of the radiation dose rate inside a person's house plus one-sixth of the outdoor radiation in his environment. The local gamma ray component is only about one-half of the total dose, the other half arising from cosmic radiation and the internal

body potassium (Spiers, F. W., personal communication, 1 May 1966)

14 See Warren, H. V. (1967) and Howe G. M. (1970)

15 Goitre has always been equally severe, if not more so, in Oxfordshire, Gloucestershire, and Dorset. It is the content of iodine in drinking water which is a determining factor in the distribution of endemic goitre

16 Crawford, M. D., Gardner, M. J., and Morris, J. N., 'Mortality and hardness of local water supplies', *Lancet*, 1 (1968), 827–83

17 Schroeder, H. A., 'Relation between mortality from cardiovascular disease and treated water supplies', *J Amer Med Ass*, 172 (17) (1960), 1902–8

18 Howe, G. M., 'The geographical distribution of cancer mortality in Wales, 1947–53', *Trans and Papers, Inst Brit Geogrs*, 28 (1960), 190–215

See also Howe, G. M., 'Disease patterns and trace elements', *Spectrum*, 77 (1970), 2–8

19 Allen-Price, E. D., 'Uneven distribution of cancer in West Devon', *Lancet*, 1 (1960), 1235–8

Chapter 4, Health Hazards of the Biological Environment (pp 40–49)

1 See note 2, Chapter 1

2 So far as communicable diseases such as smallpox, tuberculosis, and leprosy are concerned the risks to the public are well understood and cases are competently managed by the health authorities. Unfortunately this is not always so with other diseases acquired abroad which are not a public health risk in Britain but are nevertheless potentially lethal to the unfortunate victim. As stated by B. G. Macgraith (School of Tropical Medicine, University of Liverpool): 'In this jet-age a man can go half-way round the world in a fraction of the incubation period of any of the serious infectious diseases. This means that he can (and sometimes does) enter this country in the silent stage of an infection he has acquired before the clinical picture develops. He may be at home for a week or more before he gets ill and when he does, the diagnosis may be delayed or missed. . . . With the introduction of jumbo jets and supersonic airlines the situation is going to get worse' (*The Times*, 17 March 1969)

3 It is difficult to estimate the extent of malaria in Britain during early times, because the term 'ague' was applied indiscriminately to many fevers. See Smith, W. D. L., 'Malaria and the Thames', *Lancet*, 270 (1956), 433–6

4 The development of powerful chemical pesticides based on organo-phosphorus and chlorinated hydrocarbons revolutionised chemical warfare against insects and other pests which constitute a threat to the health of man and his crops. Many of these chemicals, however, may be lethal and DDT dieldrin, aldrin, and related substances persist in soils and accumulate in animal and body fat. On 28 March 1969, Sweden announced a ban on DDT and other chlorinated hydrocarbons, aldrin, dieldrin, and lindane, to take effect from 1 January 1970. The Swedish Poisons Board which motivated the ban, said that the insecticides were dangerous to animal and plant life and were long-term health risks for humans. A similar ban has been effected in the USA, and there is restricted usage in the United Kingdom

Chapter 5, Health Hazards of the Human Environment (pp 50–72)

1 Chadwick, E., *Report on the sanitary conditions of the labouring population of Great Britain, 1842*. New edition, edited and with an introduction by M. W. Flinn (Edinburgh 1965)
2 Yudkin, J. and McKenzie, J. C. (eds), *Changing food habits* (London 1964)
3 Cyclamate sweeteners were banned in Britain in October 1969
4 The Report of the Swann Committee (November 1969) recommended that the use of penicillin and the tetracyclines in animal feeding-stuffs should be prohibited and the use of antibiotics be restricted. The recommendations were accepted by the Government
5 The temperature of the air generally gets lower with increasing height but occasionally the reverse is the case, and, when the temperature *increases* with height there is said to be an 'inversion'. This leads to the formation of a high layer of warm air. Inversions of temperature are commonly experienced in hollows and valleys, especially in winter on calm, clear nights
6 Entries in Malmesbury, Earl of, *Memoirs of an Ex-Minister* (London 1885), read (for the year 1858) as follows:

'June 21st—The heat of this last month has been quite exceptional, the thermometer constantly rising to 84° . . .
June 23rd—The pestilential smell from the Thames is become intolerable, and there has been a question of changing the locality of Parliament. Nothing can be done during this heat . . .

June 27th—We have ordered large quantities of lime to be thrown into the Thames, for no works can begin until this hot weather is over. The stench is perfectly intolerable, although Madame Ristori, coming back one night from a dinner at Greenwich given by Lord Hardwicke, sniffed the air with delight, saying that it reminded her of her "dear Venice" . . .'

7 A recent Report by the Medical Research Council, *Assay of Strontium 90 in Human Bone in the United Kingdom* (HMSO, April 1969) shows a decline in all age groups far greater than was expected by a team of specialists who prepared the survey. The dramatic drop in radioactive fall-out material ingested by the population has occurred since the restriction of tests of nuclear bombs in the atmosphere

Chapter 6, Pre-Norman and Norman Times (pp 73–92)

1 Hoskins, W. G., *The making of the English landscape* (London 1963)
2 The Wall of Antonine was a turf wall erected between the Forth and Clyde; Hadrian's Wall was built along the Solway–Tyne Gap (Bowness to Wallsend). The latter defended the northern frontier of the Roman province of Britain; between it and the Wall of Antonine was a frontier zone
3 Fleure, H. J., *op cit*, 521
4 Fox, C., *The personality of Britain*, Nat Mus Wales (Cardiff 1932, 1938, 1947)
5 Hoskins, W. G., *op cit*
6 Palaeopathology deals with the excavation, treatment, and study of human skeletal remains, with particular reference to evidence of disease and abnormality in early man.
7 MacArthur, W. P., 'Some notes on old-time leprosy in England and Ireland', *J Roy Army Med Corps*, 45 (1925), 414–22
8 A leper house was endowed at Canterbury by Lanfranc, the first Norman Archbishop of that See. Others were founded later at Westminster, Southwark, Highgate, and other places in London, and there were numerous other hospitals throughout England. See Creighton, C., *op cit*, 86, and Simpson, J. Y., 'Archaeological Essays', Vol 11, 19, for a list of over 100 leper establishments in England
9 Comrie, J. D., *History of Scottish medicine to 1860*, Vol 1 (The Wellcome Historical Medical Museum, 1932), 43
10 Creighton, C., *A history of epidemics in Britain, 1894*. Second edition, with additional material by D. E. C. Eversley, E. A. Underwood, and L. Ovenall (London 1965). This great work of

scholarship is the only book which deals systematically with the history of epidemic diseases in Britain. It is an inexhaustible store of information and has been referred to frequently during the preparation of this book. Nevertheless Creighton's data and interpretations must always be handled with care. He did not believe in the germ theory of disease and some of his material was selected, if not to disprove it, as least not to support it. He believed in the localist–miasmatic theory of disease causation

11 Bonser, W., *The medical background of Anglo-Saxon England*, The Wellcome Historical Medical Library (London 1963)

12 Adamnan, *Life of St Columba*, ed W. Reeves (1874)

13 Shrewsbury, J. F. D., 'The yellow plague', *J Hist Med Allied Sci*, 4 (1949), 14

14 MacArthur, W. P., 'The identification of some pestilences recorded in the Irish Annals', *Irish Hist Studies*, 6 (1949), 172
See Bonser, *op cit*, 64–70 for discussion

15 Pelusium is now represented by two large mounds close to the coast and the edge of the desert, twenty miles east of Port Said. The easterly distributary of the Nile on which it was sited has long since been silted up

16 Procopius of Caesarea (550). *De Bello Persico*, Works of Procopius of Caesarea. Vol 2. Translated by H. B. Dewing, Loeb Classics (London). Quoted by Hirst, L. F., *The conquest of plague* (Oxford 1953), 11

17 Bede, the Venerable, *Eccles Hist*, iii, 27

18 Mullet, C. F., *The bubonic plague and England: an essay in the history of preventive medicine* (Lexington, USA 1956)

19 Comrie, J. D., *op cit*, 49

20 Creighton, C., *op cit*, 15–17

21 MacArthur, W. P., 'A brief story of English malaria', *Brit Med Bull*, 8(1) (1951), 76–9. See also Smith, W. D. L., 'Malaria and the Thames', *Lancet*, 270 (1956), 433–6

22 Creighton, C., *op cit*, 13–14

Chapter 7, Medieval Times (pp 93–106)

1 Shrewsbury, J. F. D., *A history of bubonic plague in the British Isles* (London 1970)

2 See Roberts, R. S., 'The use of literary and documentary evidence in the history of medicine' (Reference 25) in *Modern methods in the history of medicine*, ed E. Clarke (London 1971)

3 Hirst, L. F., *The conquest of plague: a study in the evolution of epidemiology* (Oxford 1953)

4 Many ports of southern England were in frequent contact with the

Continent or with the Channel Islands. The plague may have been brought from Calais or else from Jersey and Guernsey which were suffering badly from the disease at this time. See Ziegler, P., *The Black Death* (London 1968), 119–22

5 Creighton, C., *op cit*, 116

6 Ziegler, P., *op cit*, 136

7 Rees, W., 'The Black Death in England and Wales, as exhibited in manorial documents', *Proc Roy Soc Med* (*Hist Med*), 16 (1923), 27

8 Chronicle of the Scottish Nation, ed W. F. Skene (Edinburgh 1872)

9 *Cronykil of Andrew of Wyntoun*, ed D. Laing (Edinburgh 1872), Vol 11, 482

10 Rees, W., *op cit*

11 The mortality figures for Britain were paralleled on the Continent, which lost at least a fourth of its population between 1348 and 1350. Florence was reduced in population from 90,000 to 45,000 and Sienna from 42,000 to 15,000. Hamburg apparently lost about two-thirds of its inhabitants

12 Narrow, ill-cleansed streets, ditches, and streams filled with household garbage and sweepings, a virtual absence of sewage disposal, and insanitary conditions generally typified the larger towns in the fourteenth century

13 Bean, J. M. W., 'Plague, population and economic decline in the later Middle Ages', *Econ Hist Rev*, 2nd Ser, 15 (1963), 423–38

14 Thompson, J. W., 'The aftermath of the Black Death and the aftermath of the Great War', *Amer J Sociol*, 16 (1920–1), 565. See also Talbot, C. H., *Medicine in medieval England* (London 1967), 163–4 and Langer, W. L., 'The Black Death', *Scientific American*, 210, 2 (1964), 114–21. Other aspects of human behaviour in the face of the plague are considered in connection with the plague of 1665 (p 128)

15 In Bean's view (Bean, *op cit*) 'there can therefore be no doubt that when England was hit by a second plague in 1361 the population was not more than two-thirds of that of the early 1340s'

16 Langland, W. (*c* 1332–*c* 1400)—supposed English author of fourteenth-century poem *Piers Plowman*

17 Talbot, C. H., *op cit*, 162

Chapter 8, Tudor Times (pp 107–119)

1 Brown, E. H. P. and Hopkins, S. V., 'Seven centuries of the prices of consumables compared with builders' wage-rates', *Economica*, N.S., 23 (1956), 296

2 Drummond, J. C. and Wilbraham, A., *The Englishman's food. A history of five centuries of English diet*. (Revised by D. Hollingsworth)

3 The effects of the climatic deterioration from about AD 1300 to 1500-1600, and after have commonly been ignored, presumably becaus they were heavily overlaid by the Black Death of 1349–50 and th recurring ravages of the plague thereafter. See Lamb, H. H., *op ci*

4 Saltmarsh, J., 'Plague and economic decline in the later Middl Ages', *Camb Hist J*, 7 (1941), 23–41

5 The Company of Parish Clerks was granted a charter in th thirteenth century. At this time its function was connected wit church music

6 Graunt, J., *Natural and political observations . . . made upon the Bills o Mortality* (London 1662)

7 Caius, J., *A Boke or Conseill against the disease called the sweate, o sweatying sicknesse* (London 1552)

8 The epidemic of syphilis in England was used to discredit th Romish mass-priest, the monasteries, and orders of friars. Th scandalous lives of priests, monks, and friars made the stronges argument for the policy that Henry VIII had adopted against papa supremacy and in favour of Reformation

9 Creighton, C., *op cit*, 415

10 Hirsch, A., *Handbook of geographical and historical pathology* (1881, ir German). English translation by C. Creighton, Sydenham Society 1883–6, 92–8

11 Clowes, W., *A short and profitable treatise touching the cure of th disease called morbus gallicus by unctions* (John Daye, London 1579)

12 Shrewsbury, J. F. D., 'Henry VIII: A medical study', *J Hist Me Allied Sci*, 7(2) (1952), 141–85

13 Hippocrates (460–*c* 360 BC) and Hippocratic writers gave genera descriptions of the weather and climate associated with variou diseases prevalent in the island of Thasos, off the coast of Thrace Atmospheric conditions were grouped under four constitution (katastases). The first constitution, marked on the whole by cool dry weather, was related mainly to mumps and pulmonary illness The humid and warm fourth constitution was one described as 'pestilential' by Galen (AD 130–200). Galen developed the idea o miasmatic corruption of the air

14 Drummond, J. C., *et al*, *op cit*, 133

Chapter 9, Stuart Times (pp 120–136)

1 Pepys, S., *Pepys' Diary*, ed J. P. Kenyon (London 1963)

2 Defoe, D., *A journal of the plague year, 1722*. Introduction by D. J. Johnson (London 1966)

3 Hirst, L. F., *The conquest of plague: a study in the evolution of epidemiology* (Oxford 1953), 53

4 Mead, R., *A discourse on the plague* (London 1720), I, 290
5 Quoted from Batho, C. R., 'The plague of Eyam: a tercentenary re-evaluation', *Derbyshire Archaeol J*, 84 (1964), 88–9. On p 86 he writes: 'It is just conceivable that the Plague of Eyam had its origin in Derby. The village wakes, held on the first Sunday after the festival of St Helen to whom Eyam Church was dedicated (18 August), are said by the oral tradition to have been visited by a larger number than usual in 1665. The Victorians speculated that it was premonition of impending doom for the village which brought the crowds; a more likely explanation is that it was the unusually fine weather. . . . One of these visitors might have been a friend or relative from Derby, unwittingly transporting a flea bearing the fatal bacillus'
6 Creighton, C., *op cit*, 1, 684
7 Quoted by Ferguson, T., *The dawn of Scottish social welfare* (London 1948), 106
8 Langer, W. L., *op cit*, 114–21
9 In an epidemic in 1563 Queen Elizabeth took refuge in Windsor Castle and had a gallows erected on which to hang any one who had the temerity to go to Windsor from plague-ridden London
10 See Opie, I. and Opie, P., *Oxford dictionary of nursery rhymes* (1957)
11 Thomson, G., *Loimotamia, or the pest anatomized* (London 1666). Quoted by Hirst, L. F., *op cit*, 2, who also cites several other remedies and religious and magical measures resorted to in an attempt to cope with plague
12 *Mycobacterium tuberculosis* was first recognised by Robert Koch in 1882
13 Brownlea, J., 'The history of the birth and death rates in England and Wales taken as a whole from 1570 to the present time', *Pub Health* 29 (1916)
14 It was the epidemic of 1670 which gave Thomas Sydenham (1624–89), the 'English Hippocrates', the opportunity to write his classical description of measles
15 Harris, W., *Tracatus de morbis acutis infantum* (London 1689). English translation by Cockburn (1693). Quoted by Creighton, C., *op cit*, 750
16 Creighton, C., *op cit*, Vol 11, 313–39
17 Oliver Cromwell is said to have died of influenza in 1658
18 Willis, T., *Diatribae dual* (The Hague 1659)
19 Cited by Creighton, C., *op cit*, Vol 11, 338
20 Evelyn, J., *Fumifugium, or the inconvenience of the aer and smoake of London dissipated* (London 1661), vii
21 A certain Mr Corbyn Morris
22 Rogers, T. J. E., *The history of agriculture and prices in England*, Vol

V (1866–7). As illustration of the last of the periods mentioned by Rogers the following entries have been extracted from 'Calenda of Historic Weather Events since 1500', in Lamb, H. H. (1964) *op cit*, Appendix 2:

21 April 1695	'Strong SW winds and rain ended exception ally long severe and snowy winter, almost the first rain for several months near London NE and E winds had been almost continuous since 21 March . . .' (This April and May the greatest extension of the Arctic sea ice ever known was spreading round the entire coast of Iceland.)
21 August 1695	North winds and night frost at the end of a cold summer with continual rain and W gales. 'Greater frosts were not always seen in winter' (John Evelyn at Wotton, Surrey). (This summer was one of the first of a sequence of disastrous harvests in Scotland, where famine ensued.)
18 August 1696	End of the rains in the South, where W winds brought mostly fair weather over the next month: dearth of food becoming serious in Scotland.
11 December 1696	E wind brought in spell of snowy weather lasting till February, 1697.
7 June 1697	Uncommonly wet spring in England and Ireland with frequent rain and hail.
14 May 1698	Weather mostly poor till 20 August.
19 May 1698	Deep snow in Shropshire.

23 Drummond, J. C. and Wilbraham, A., *op cit*, 101

24 Mentioned by Ferguson, T., *op cit*, 110

Chapter 10, Early Industrial Times (pp 137–154)

1 Sufferers from gout have large quantities of uric acid circulating in their blood streams. Gout and uric acid levels are controlled by several factors of which the social class factors responsible for gout are quite distinct from those which affect the level of uric acid. See Acherson, R. M., 'Social class gradients and serum uric acid in males and females', *Brit Med J*, 4 (1969)

2 Dr (Rev) Alexander Webster was minister of the Tolbooth Church of Edinburgh

3 Three possible causes of a reduction in the death rate from in-

fectious diseases are (1) specific medical therapy, (2) changes in the balance between the virulence of the infective organism and its host, (3) improvements in the environment. These causes are examined by T. McKeown and R. G. Brown ('Medical evidence related to English population changes in the eighteenth century', *Population Studies*, 9(2) (1955), 119–41) who conclude that the improvement resulted from an improvement in economic and social conditions. See also Habakkuk, H. J., 'English population in the eighteenth century', *Econ Hist Rev*, 2nd Ser, 6 (1953), 117–33 and Griffith, G. T., *Population problems of the age of Malthus* (Cambridge 1926)

4 The Kirk Records of Aberdeen, quoted by Thomas Ferguson, in *The dawn of Scottish social welfare* (London 1948)

5 E. Ashworth Underwood (one-time Director, the Wellcome Historical Museum, London) quoted in a Park-Davis publication, *Jenner: Smallpox is Stemmed* (1966)

6 Typhus and typhoid were listed together in the General Register Office up to 1869. Synonyms for typhus (tuphos = fog) included epidemic fever, spotted fever, putrid fever, gaol fever, camp fever, fourteen-days fever, malignant fever, and petechial fever

7 It was possible to hold a 'frost fair' and carnival on the frozen River Thames. There were six such occasions in the 1700s

8 Generally through imported foods, as at Bournemouth, Poole and Christchurch in 1936, Croydon in 1937, and Aberdeen in 1964

9 Fothergill, J., *An account of the sore throat attended with ulcers* (London 1748)

10 Huxham, J., *An essay on fevers* (London 1750)

11 The winter of 1739–40 was one of intense frost (see note 7) and the beginning of a two-year sickly period in which typhus and dysentery reached heights unprecedented in the eighteenth century. See Creighton, C., *op cit*, Vol 11, 693

12 Creighton, C., *op cit*, Vol 11, 645

13 Drummond, J. C. and Wilbraham, A., *op cit*, 244. Creighton considers that diarrhoea of infants in London was probably at its worse in the period 1720–40 during the era of cheap gin. This opinion may have been influenced by the fact that Creighton himself was teetotal

14 Griffith, G. T., *op cit*

15 McKeown, T. and Brown, R. G., *op cit*

Chapter 11, Early Victorian Times (pp 155–183)

1 The end of the Salt Tax made possible the commercial production of synthetic soda which ended the burning of seaweed to produce

kelp on the coasts of the Islands and Highlands of Scotland, an industry which had supported a larger population than was possible by agriculture alone

2 See Elkins, T. H., 'National characteristics of industrial landscapes', *Mélanges de Géographie* (Gembloux 1967)

3 Gaskell, P., *The manufacturing population of England: its moral, social and physical conditions and the changes which have arisen from the use of the steam machinery* (London 1833)

4 Kay, J. P., *The moral and physical condition of the working classes employed in the cotton manufacture in Manchester* (Manchester 1832)

5 *Parliamentary Papers*, Vol XVIII (1845)

6 From a Report on the State of the Public Health in Birmingham by a Committee of Physicians and Surgeons. *Parliamentary Papers*, Vol XIV (1843)

7 Report on the Sanitary Condition of Bradford by James Smith of Deanston. *Parliamentary Papers*, Vol XVIII, 2, 'Appendix' (1895)

8 Symons, J. C., *Parliamentary Papers*, Vol XIV, E.11 (1843)

9 Report by James Smith, *Parliamentary Papers*, Vol XVIII (1845)

10 Report by J. R. Martin, *Parliamentary Papers*, Vol XVIII (1845)

11 Report by the Rev Whitwell Elwin, *Parliamentary Papers*, Vol XXVI (1842)

12 Report by Dr Laurie, *Parliamentary Papers*, Lords, Vol XXVI (1842)

13 In Reports from Assistant Hand-Loom Weavers' Commissioners for the South of Scotland by J. C. Symons. *Parliamentary Papers*, Vol XLII (1839)

14 Cobbet, William, Rural Rides (for 7 November 1821), 1830 (1835 edition)

15 Guy, W. A., First Report of the Health of Towns Commissioners, Vol 1, p 82 (1835 edition)

16 Gaskell, P., *op cit*

17 Frederick (Frederich) Accum (1769–1838) exposed the adulteration of food in his *Treatise on the Adulteration of Food and Culinary Poisons* (London 1820). Accum became the subject of a bitter campaign of abuse on the part of those whom he had ruthlessly exposed. Unfortunately he himself provided them with the opportunity for revenge. Seemingly he tore pages from certain books in the library of the Royal Institution in London which obtained a bill of indictment against him. Before the case was heard Accum fled to Germany and forfeited £200 bail

18 Chadwick, E., *op cit*, 220

19 Previous to July 1837, no distinction was made in published tables between typhus fever and enteric fever

20 Chalmers, A. K., *The health of Glasgow, 1818–1925: an outline* (Glasgow 1930), 2–3
21 Chalmers, A. K., *op cit*, 4
22 Currie, J., *Medical reports* (Liverpool 1797), 204
23 Shapter, T., *The history of the cholera in Exeter in 1832* (London 1841)
24 See Jones, G. P., 'Cholera in Wales', *Nat Lib Wales J*, 10 (1957–8), 281–99
25 First Report of the Registrar-General, Appendix I (London 1839)
26 Brownlea, J., 'An investigation into the epidemiology of phthisis in Great Britain and Ireland', *Med Res Council, Special Reports Series*, No 18, Table XXV (1918)
27 The range of earnings among wage-earners in the early nineteenth century was so large that it is impossible to speak of a single 'working class'
28 Gaskell, P., *op cit*
29 Burnett, J., *Plenty and want. A social history of diet in England from 1815 to the present day* (London and Edinburgh 1966)
30 Dickens, C., *Oliver Twist* (London 1838). *Oliver Twist* and *Bleak House* bear the impress of the close relationship between Dickens and Dr Southwood Smith (see p 186), the Unitarian physician who devoted his life to the betterment of the poor, especially of the sick poor. In their pages occur passages which are taken almost verbatim from one of Smith's reports to Parliament
31 Hobson, W., *World health and history* (Bristol 1963). He cites Keats, Shelley, Stevenson, *et al*
32 Ferguson, T. *et al*, 'Public health in Britain in the climate of the nineteenth century', *Public health and urban growth*, Report No 4. Centre for Urban Studies, University College (London 1964)
33 Prothero, R. E., speaking of the standard of life from 1800 to 1834 in *The pioneers and progress of English farming* (London 1888)

Chapter 12, Late Victorian Times (pp 184–199)

1 See Finer, S. E., *The life and times of Sir Edwin Chadwick* (1952) also Lewis, R. A., *Edwin Chadwick and the Public Health Movement, 1832–48* (London 1952)
2 Greenwood, M., *Epidemics and crowd diseases: an introduction to the study of epidemiology* (London 1935)
3 Badham, C., *Observations on the inflammatory affections of the mucous membranes of the bronchia* (London 1808)
4 Although it was first recognised and described as a sharply defined disease by P. Bretonneau in 1826, the term 'diphtheria' was first introduced by B. Godfrey in 1857. It is of interest in the context of disease classification to recalled the contribution of Miss Florence

Nightingale. Cecil Woodham-Smith reminds us that in 1859 each hospital followed its own method of naming and classifying diseases and that Miss Nightingale embarked on a campaign for uniform hospital statistics. With the help of Sir James Page and Sir James Clarke she drew up a standard list of diseases and drafted model hospital statistical forms which would, she wrote, 'enable us to ascertain the relative mortality of different hospitals, as well as of different diseases and injuries at the same and at different ages, the relative frequency of different diseases and injuries among the classes which enter hospital in different countries and in different districts of the same countries' (C. Woodham-Smith, *Florence Nightingale, 1820–1910*, 335)

5 Creighton, C., *op cit*, Vol 11, 739
6 Benjamin, B., 'Urban background to public health changes', *Public health and urban growth*, by T. Ferguson *et al*, Centre for Urban Studies, University College, Report No 4 (London 1964), 16
7 Keats, J., *Ode to a Nightingale*
8 Drummond, J. C. and Wilbraham, A., *op cit*, 403
9 Burnett, J., *op cit*, 214
10 Report of the Inter-Departmental Committee on Physical Deterioration (London 1904)
11 Rowntree, S. B., *Poverty: a study of town life* (London 1901)
12 Booth, C., *Life and labour of the people of London* 17 vols (London 1902)
13 McKeown, T. and Brown, R. G., *op cit*, 1955. McKeown, T. and Record, R. G., 'Reasons for the decline of mortality in England and Wales during the nineteenth century', *Population Studies*, 16 (1962), 94–122. McKeown, T., 'Medical issues in historical demography', in E. Clarke (ed), *op cit*

Chapter 13, Modern Times—Morbidity (pp 200–220)

1 Typhoid in Britain used to be pre-eminently a water-borne disease but since the great improvement during the present century in the purification of drinking water this method of carriage in this country has become rare. The place of water has been usurped by foods, mainly imported
2 There have been some classic carriers in medical history. Perhaps the most famous was 'Typhoid Mary' in the USA. She was Mary Mallon, a New York cook, who contracted typhoid in 1901, infected various families for whom she worked and probably caused an epidemic in New York State in 1903 with 1,300 cases
3 Tanner, J. W., 'Earlier maturation in man', *Scientific American*, 218 (1968), 21

4 Office of Health Economics, *Pneumonia in decline* (London 1963)
5 Office of Health Economics, *The price of poliomyelitis* (London 1963)
6 Office of Health Economics, *The lives of our children: a study in childhood mortality* (London 1962)
7 Figs 69–76 are adapted, with permission, from maps in *Reader's Digest Atlas*
8 Alcoholism is not given as a reason for absence from work but recent reports (*A survey of alcoholism in an English county* by M. C. Moss and Beresford Davies (1968) and *A study of the prevalence, distribution and effects of alcoholism in Cambridgeshire* by Griffith Edwards) implicate this 'widespread, killing, largely preventable disease which has a devastating effect on family life and is causing heavy losses to industry every year' in frequent lateness at work and Monday morning absenteeism
9 Pregnancy (uncomplicated) comes into the 'non-sickness' category; so do preventive injections

Chapter 14, Modern Times—Mortality (pp 223–244)

1 Figs 79–82, 84–87, and 90 are simplified versions of maps which appear in *A national atlas of disease mortality in the United Kingdom* by G. M. Howe, Second (enlarged and revised) edition (Nelson, London 1970)
2 Demographic base maps, in which the areas of the 'squares' and 'diamonds' are proportional to the populations of the places they represent, are used for the presentation of standardised mortality ratios in preference to geographical base maps since they relate the SMRs to the local 'population at risk' rather than to the areas in which the people live
3 The index used for mapping is the 'standardised mortality ratio' (SMR). This ratio makes due allowance for differences in the local age structure of populations by a process of 'indirect standardisation'. The SMR provides a means of comparing mortality in any local area with that of Britain as a whole and is a more reliable index than is the crude death rate
4 Howe, G. M., 'The geography of death', *New Scientist*, 38 (1968), 612–14
5 Todd, G. F. (Ed.), 'Statistics of smoking in the United Kingdom', Research Paper No 1, 4th ed (1966)

Glossary of Medical Terms

(to be used in conjunction with the Index)

Some of the terms are not, or only partially, explained in the text.

Aetiology (Etiology) The science of the causes of disease.

African Sleeping Sickness See **Trypanosomiasis.**

'Ague' Sometimes used as synonymous with malaria although not all agues were malarial. The term ague originally meant any acute fever and later was often applied specifically to typhus ('burning ague').

Angina Pectoris A sudden and violent pain in the chest which is usually precipitated by exertion in patients with diseased coronary arteries.

Anopheles Genus of mosquitoes, several species of which transmit the agents of malaria, the so-called *Plasmodia.*

Antibiotics Substances extracted from fungi which kill or inhibit the growth of organisms causing disease.

Arthritis Any disease or disability in which there is degenerative change in a joint.

Asbestosis An occupational disease of the lungs which may develop in those exposed to inhalation of asbestos fibres.

Asthma An allergic lung condition associated with shortness of breath. If the condition is the result of obstructed airways it is said to be *bronchial asthma.* If red blood cells fail to carry oxygen away from the lungs it is *cardiac asthma.*

Atherosclerosis (Atheroma) A disease of the innermost coat (intima) of the aorta and of its main branches. The disease is characterised by nodular patchy fatty degeneration resembling porridge in appearance (athêrê, porridge).

Athlete's Foot (*Tinea pedis*) A fungal infection of the feet.

BCG Vaccination (in tuberculosis) Consists of a live though harmless bacillus originally produced by Calmette and Guerin in 1906 (hence the initials BCG). The strain is mild and does not produce active tuberculosis but it establishes resistance.

Bacteria Unicellular organisms. They are among the smaller living creatures known and because of their minute size are often termed micro-organisms.

Botulism A serious type of food poisoning caused by a toxin produced by the micro-organism *Clostridium botulinum.*

Bronchitis Essentially inflammation of the mucous membrane which lines the bronchi leading from the windpipe to the lungs and down which passes the air in breathing. *Acute Bronchitis* refers to a brief attack of inflammation. *Chronic Bronchitis* refers to long-standing inflammation of the bronchi and is often associated with fibrosis, emphysema, asthma, and chronic sinusitis.

Broncho-Pneumonia See **Pneumonia.**

Cancer (Malignant Tumour, Malignant Neoplasm) An abnormal proliferation of new growth in cells and tissues that produces harmful and often fatal effects. The basic cause, ie the 'carcinogens', is unknown. Common cancers include cancer of the skin, cancer of the lung, cancer of the stomach, cancer of the breast, cancer of the pancreas, cancer of the prostate, and cancer of the cervix.

Cancer of the Lung and Bronchus See **Cancer.** The aetiology of lung-bronchus cancer, as indeed of any other cancer, is not yet understood but the causal relationship with smoking is now well established.

Cardio-Vascular Disease Disease associated with the heart and arteries.

Cerebrospinal Fever (Epidemic Meningitis) See **Meningitis.**

Cerebro-Vascular Disease ('Stroke', 'Shock') Haemorrhage or softening of the brain.

Chemotherapy The cure of disease by the administration of chemical substances which kill the organisms causing it without at the same time damaging the body.

Chloramphenicol (Chloramycetin) An antibiotic substance used in the treatment of many infectious diseases.

Cholera (Asiatic Cholera) A disease caused by the *Comma bacillus* (*Vibrio cholerae*) discovered in 1883 by Koch, and its toxins. The bacilli are ingested in contaminated water and food and lead to diarrhoea, vomiting, and dehydration of the body.

Cholesterol One of the lipids found in bile, gallstones, brain, blood cells, plasma, egg yolk, seeds and animal tissues, etc.

Consumption See **Pulmonary Tuberculosis.**

'Continued Fever' ('Putrid Malignant Fever') See **Typhus** and **Typhoid.**

'Convulsions' See **Infantile Diarrhoea.**

Coronary Thrombosis The coagulation of blood during life in one of the coronary arteries.

Dental Caries Decay of the teeth.

Derbyshire Neck See **Goitre.**

Diabetes Mellitus A disease associated either with a breakdown of the body's own supply of insulin through damage to the pancreas or with a change in the sensitivity of the body's response to insulin.

Diphtheria ('Croup') Also known as malignant angina—is a throat infection with possible fatal outcome. It is caused by the diphtheria bacillus and its chemical toxin.

Dysentery An acute form of diarrhoea with spasms in the bowels and bloody excretion, called 'bloody flux' by our forebears. It is caused by either the dysentery bacilli (bacillary dysentery) or the dysentery amoebas (amoebic dysentery).

Emphysema A respiratory disease associated with overdistention of the lungs.

'English Sweat' ('Sweating Sickness') A mysterious and deadly disease which affected England in 1486, 1507, 1518, 1529, and 1551. See also **Influenza.**

Enteric Fever ('Slow Nervous Fever', 'Continued Fever') See **Typhoid** and **Paratyphoid.**

Epidemiology The study of the spread of diseases.

Erysipelas (St Anthony's Fire) An acute bacterial infection of the skin, usually of the face, in which the affected parts are of a deep red colour.

Escherichia coli Bacterium, normally resident in the colon, of which only certain types are pathogenic.

'Farmer's Lung' ('Allergic Alveolitis') A disease of the lungs resulting from the handling of a variety of dusty, mouldy, organic materials such as mouldy hay. The pulmonary condition results from the inhalation of fungal spores, especially of actinomycetes (*Thermopolyspora polyspora*).

Gastric and Duodenal Ulcer (Peptic Ulcer) Localised defect in the lining of the stomach or duodenum.

Gastro-Enteritis Inflammation of the stomach and of the intestines usually produced by some of the Salmonella group of bacteria.

Goitre (Derbyshire Neck) A thyroid enlargement, thought to be due to, or closely linked with, iodine deficiency.

'Griping of the Guts' See **Infantile Diarrhoea.**

Haemoptysis The spitting up of blood.

Hay Fever (Allergic Rhinitis) Disease of the nose associated with any allergen such as dust, rye-grass, or ragweed and characterised by impaired nasal breathing. Symptoms are similar to those for a head cold (common cold).

Hookworm Disease (Ancyclostomiasis) A disease of tropical climates, but may be encountered in cooler regions, particularly in mines and tunnels. It is caused by a parasitic roundworm and causes severe anaemia, debilitation, and lowered efficiency.

Hypertension High blood pressure.

Hypothermia Lowered body temperature.

Infantile Diarrhoea ('Summer Diarrhoea', 'Convulsions') A definite epidemiological entity, due either to a bacterium or a virus, which reached its zenith in Britain in the later years of the nineteenth century. It was then a nationwide scourge in every hot summer.

Infective Hepatitis (Catarrhal Jaundice) Inflammation of the liver cells caused by a virus.

Influenza ('New Disease', 'Hot Ague', 'New Ague', 'New Fever', 'New Pestilence') An acute infectious respiratory disease caused by a virus. Whether the so-called 'Sweating Sickness' (*q.v.*) that swooped on Britain in the fifteenth century and the first half of the sixteenth century was a form of influenza or a distinct disease, remains an unsolved mystery.

Ischaemic Heart Disease A heart disease resulting from a decreased blood supply to the heart. This may be due to narrowing or occlusion of the coronary arteries.

Leprosy (Hansen's Disease) A chronic infectious disease of the skin and nerves causing mutilations and deformities. The disease called 'zara'ath' in the Bible, usually translated as 'leprosy', is a generic term referring probably to several skin diseases of which true leprosy (caused by *Mycobacterium leprae*) may be one.

Lipids A general term that includes fats and fat-like compounds.

Malaria ('Lencten Adl', 'Spring Ill') Now thought of as a 'hot country' (tropical) disease but it was at one time endemic in certain marshy districts of Britain such as the Fens, the Isle of Sheppey, south-east Kent, and Somerset Levels. It is due to a group of protozoa, with complicated life histories, which are carried from infected to uninfected persons by amopheline mosquitoes. The word 'ague' (*q.v.*) once a popular name for malaria, was merely an acute fever rather than an intermittent or paroxysmal one.

Measles (Morbilli, 'Mezils') A disease caused by a virus. In the Middle Ages the term 'measles' was used in a generic sense to designate every chronic skin disease and it was not until the sixteenth century that measles was distinguished from smallpox and/or scarlet fever.

Meningitis Inflammation of the meninges. It is of several forms and occurs usually in the wake of other general infections, eg tuberculosis or septicaemia caused by streptococci or staphylococci. Only meningococci meningitis (cerebrospinal fever, spotted fever) occurs in epidemic form.

Metabolism An inclusive term which applies to virtually all the active processes in living organisms.

Myocardial (Cardiac) Infarction See **Coronary Thrombosis.**

Osteo-Arthritis Degenerative joint disease is the preferred term for this disease because it describes the underlying degenerative process in contrast to the inflammatory process which accompanies rheumatoid arthritis (*q.v.*).

Pandemic An epidemic spreading all over the world.

Paratyphoid Fever See **Typhoid** and **Paratyphoid.**

Penicillin The first antibiotic, discovered by Fleming, which revolutionised the treatment of many infectious diseases having a bacterial aetiology.

Peptic Ulcer See **Gastric** and **Duodenal Ulcer.**

Pernicious Anaemia A nutritional deficiency disease. The basic defect is the absence of a gastric intrinsic factor without which vitamin B_{12} cannot be absorbed.

Pestilence A word used in a generic sense much like 'the plague', 'fever', 'measles', etc. It referred to any kind of acute epidemic. The bubonic plague which invaded Britain in the fourteenth century was called 'The Great Pestilence' by contemporary chroniclers.

Phthisis See **Pulmonary Tuberculosis.**

Pituitary Adenoma A simple tumour of the pituitary gland. This gland controls the thyroid, adrenals, ovaries, etc, and an adenoma may stimulate these to excessive activity.

Plague A term formerly used loosely to apply to every epidemic resulting in high mortality. Later used exclusively to define the particular epidemic which, recognised by specific clinical, epidemiological, and bacteriological characteristics, is now called plague. There are three forms of plague: (*a*) bubonic plague ('botch'), with such characteristic symptoms as swelling of lymph nodes in the groins and armpits, (*b*) pneumonic (pulmonary) plague in which the victims appear blue-black, and (*c*) septicaemic plague in which the organisms proliferate in the blood stream.

Pneumoconiosis Lung disease due to inhalation and retention in the lungs of industrial dusts, with the production of fibrosis.

Pneumonia There are two types, Lobar pneumonia (pneumococcal) and Broncho-pneumonia. *Lobar pneumonia* is due to infection with pneumococci and is a severe, occasionally fatal disease if untreated. The disease responds dramatically to sulphonamides, penicillin, and other antibiotics. *Broncho-pneumonia* is characterised by a gradual spread of infection from the bronchi to the lungs. The infection may be caused by a variety of organisms and the response to therapy is less dramatic.

Poliomyelitis A febrile disease affecting mainly the spinal chord of the central nervous system. It carries the risk of permanent paralysis. It is caused by three types of virus referred to as Types, I, II, and III.

Psycho-Neurotic Disorders Disorders associated with anxieties of one kind or another.

Psychoses A group of serious mental disorders characterised by primary disturbance of emotional feeling, acute depression, mania, delusions of persecution, etc.

'Quinsy' Severe inflammation of the throat or tonsils, usually following tonsilitis.

Relapsing Fever Caused by *Spirochetes*, is characterised by anything between two and ten relapses. It is transmitted by human head lice (*Pediculus humanis corporis*) and by ticks.

Rheumatic Heart Disease (Rheumatic Fever) A generalised disease manifest by inflammation of the heart but mainly affecting the valves. The exact cause of rheumatic heart disease is not yet known.

Rheumatism A general descriptive term meaning discomfort, pain, and stiffness in or around muscles and joints. The term **Arthritis** (*q.v.*) is used when there is pathological change in a joint.

Rheumatoid Arthritis A chronic, generally progressive disease, involving inflammation of several joints.

Rhinitis Acute and chronic diseases of the nose characterised chiefly by impaired nasal breathing. In its acute form, such as a head cold, rhinitis is accompanied by a profuse watery discharge.

Rickets (Rachitis) A deficiency disease which is invariably due to a shortage of vitamin D in the body. This may result from inadequate intake of this vitamin or lack of sunshine.

Rickettsiae Minute micro-organisms which resemble very small bacteria or large viruses. Some cause important diseases in man such as Typhus (*q.v.*), Trench fever, and Q. fever.

St Vitus's Dance (Chorea) A nervous manifestation of that process which usually results in rheumatic heart disease (*q.v.*).

Scarlet Fever (Scarlatina) A reaction to a streptococcol infection. It is characterised by a throat infection, high temperature, and a red rash.

Scurvy A deficiency disease caused chiefly by the lack of fresh vegetables and fruit, more particularly of the vitamin C (ascorbic acid) they contain.

Sibbens See **Syphilis.**

Silicosis A severe lung disease which occurs as a result of inhalation of fine particles of quartz and particles of silicon-containing rock.

Smallpox (Variola, 'The Pox') A deadly and contagious disease caused by a virus. It was widespread and common in eighteenth-century Britain but diminished following the introduction of vaccination.

Spirochaetes A small corkscrew-like organism which correspond more closely to protozoa than bacteria.

Sporozoite The form of the malaria parasite in which it is transferred to man by the mosquito.

Streptococcus A micro-organism which is round and arranged in chains. It may cause such conditions as acute tonsillitis, erysipelas, etc.

Syphilis ('Great Pox' or 'French Pox') One of the major venereal diseases. The causative organism—a spirochete, the *treponema pallidum*—is usually acquired during sexual intercourse with an infected person. Such diseases as pinta and yaws are caused by similar organisms.

Tetanus (Lockjaw) A disease characterised by muscular rigidity and spasms and caused by the bacillus *B. tetani* and its exotoxin.

Toxaemia of Pregnancy An undesirable complication of pregnancy associated with high blood pressure, accumulation of fluid, and disturbances of the urinary and nervous systems.

Trench Fever A rickettsial disease caused by *Rickettsia quintana* and transmitted from person to person by the body louse. Symptoms include rapid onset of fever, accompanied by headache, varying degrees of mental clouding, prostration, and rash.

Trypanosomiasis (African Sleeping Sickness) A disease of the blood stream, lymph nodes, and central system caused by *Trypanosoma gambiense* carried by the tsetse fly *Glossina palpalis.* The disease is characterised by, among other things, profound lethargy and hence the name 'sleeping sickness'.

Tuberculosis (Consumption, Phthisis) The result of an infection of various organs with a bacillus known as the tubercle bacillus, *Myobacterium tuberculosis.* Many different organs can be affected but disease of the lungs is by far the most common form.

Typhoid (Enteric) and Paratyphoid Fever An acute infectious disease of man caused by a bacterium *Salmonella typhosa.* Paratyphoid fever closely resembles typhoid though usually milder, caused by *Salmonella* organisms of other species.

Typhus ('Gaol Fever', 'Ship's Fever', 'Hospital Fever', 'Synochus') A disease of famines and was characterised by high fever, headache, numbness, and by the so-called petechiae, ie red spots on the skin resembling flea bites (hence 'petechial typhus'). In its classic form it is caused by *Rickettsia prowazekii* and is transmitted by lice from man to man.

Venereal Diseases A group of infections which have in common the same means of transmission. The causative organisms are usually acquired during sexual intercourse with an infected person. The major venereal diseases are syphilis (*q.v.*) and gonorrhoea.

Viruses Self-reproducing agents which are smaller than bacteria and which multiply only within living cells. They are responsible or potentially responsible for a wide range of infectious diseases including smallpox, measles, and influenza.

Whooping Cough (Chin-Cough, Pertussis) An acute, highly communicable respiratory disease characterised in its venal form by bouts of coughing, followed by a long-drawn inhalation or 'whoop'. It is caused by bacilli, *Haemophilus pertussis*.

Yellow Fever An acute infectious disease of tropical and subtropical regions which is capable of invading the temperate zones. It is caused by a virus transmitted by a domestic mosquito *Aedes aegyptii*.

Bibliography

Abbott, D. C. and Thomson, J., 'Pesticide residue analysis', *World Rev Pest Control*, 7(2) (1968), 70–83

Accum, Frederick (Frederich), *Treatise on the adulteration of foods and culinary poisons* (London 1820)

Acherson, R. M., 'Social class gradients and serum uric acid in males and females', *Brit Med J*, 4 (1969), 65–7

Acland, H. W., *Memoir of the cholera at Oxford in the year 1854* (London 1856)

Allen-Price, E. D., 'Uneven distribution of cancer in West Devon', *Lancet*, 1 (1960), 1235–8

Arbuthnot, J., *An essay concerning the effects of air on human bodies* (London 1733)

Ashby, Sir Eric, *Royal Commission on Environmental Pollution. First Report, 1971* (HMSO)

Badham, C., *Observations on the inflamatory affections of the mucous membranes of the bronchia* (London 1808)

Baker, R., *Report of the Leeds Board of Health* (Leeds 1833)

Banks, A. L., 'The study of the geography of disease', *Geogr J*, 125(2) (1959), 199–216

Barbe, L., 'The plague in Scotland', *Chambers's J*, 4 (1913–14), 284–6

Barker, T. C., McKenzie, J. C., and Yudkin, J., *Our changing fare: 200 years of British food habits* (London 1966)

Bartholomew, J. G. (ed), *Bartholomew's Gazetteer of the British Isles* (Edinburgh 1902)

Batho, C. R., 'The plague of Eyam: a tercentenary re-evaluation', *Derbyshire Archaeol J*, 84 (1964), 81–91

Bean, J. M. W., 'Plague, population and economic decline in the later Middle Ages', *Econ Hist Rev*, 2nd Ser, 15 (1963), 423–38

Bede, the Venerable, *A history of the English Church and People*. Translated, and with an introduction, by Leo Sherley-Price (Harmondsworth 1956)

Belasco, J. E., 'Characteristics of air masses over the British Isles', *Geophys Mem*, 87, HMSO (London 1952)

Bell, W. G., *The Great Plague of London in 1665*. Rev ed (London 1951)

Bonser, W., 'Epidemics during the Anglo-Saxon period', *J Brit Archaeol Ass*, 3rd Ser, 9 (1944), 48–71

—— *The medical background of Anglo-Saxon England* (Publications of the Wellcome Historical Medical Library, New Ser No 3) (London 1963)

Booth, C., *Life and labour of the people of London* (London 1889–97)

Boyd, J. T., 'Climate, air pollution and mortality', *Brit J Soc Prev Med*, 14(3 (1960), 123–35

Brash, J. C., 'The Anglo-Saxon cemetery at Bidford-on-Avon, Warwickshire Notes on the cranial and other skeletal characters', *Archaeologia*, 7 (1923), 106

Brockington, F., *World Health* (Harmondsworth 1958)

Brooks, C. E. P., *The English Climate* (London 1954)

Brothwell, D. R., 'The palaeopathology of early British man: an essay on th problems of diagnosis and analysis', *J Roy Anthrop Inst*, 91(2) (1961) 318–44

—— *Digging up bones* (Brit Museum, Nat Hist) (London 1963)

Brothwell, D. R. and Sandison, A. T. (eds), *Diseases in antiquity* (Springfield Illinois 1967)

Brown, E. H. P. and Hopkins, S. V., 'Seven centuries of the prices of con sumables compared with builders' wage-rates', *Economica*, N.S., 2 (1956), 296

Brown, E. S., 'Distribution of the ABO and rhesus (D) blood groups in th North of Scotland', *Heredity*, 20 (1965), 289–303

Brownlea, J., 'The history of the birth and death rates in England and Wale taken as a whole from 1570 to the present time', *Pub Health*, 29 (1916) 211–22 and 228–38

—— 'An investigation into the epidemiology of phthisis in Great Britain an Ireland', *Med Res Council, Special Reports Series*, 18 (1918)

Brunt, D., 'Some physical aspects of the heat balance of the human body' *Proc Phys Soc*, 59 (1947), 713–26

Buchanan, A. and Mitchell, A., 'The influence of the weather on mortalit from different diseases and at different ages', *J Scot Met Soc*, 4 (1873–5) 187–265

Bulloch, W., *The history of bacteriology* (London 1938)

Burnett, J., *Plenty and want. A social history of diet in England from 1815 to th present day* (London and Edinburgh 1966)

Burnett, M., *Natural history of infectious disease* (London 1962)

Caius, J., *A Boke or Counseill against the disease called the sweate, or sweatyin sicknesse* (London 1552)

Calder, R., *The life savers* (London 1961)

Carpentier, E., 'Autour de la peste noire: famines et epidemies dans l'histoire du XIVe siecle', *Annales Economies Societés Civilisations*, 17 (1962) 1062–92

Chadwick, E., *Report on the sanitary condition of the labouring population of Grea Britain, 1842*. Edited, and with an introduction, by M. W. Flin (Edinburgh 1965)

Chalmers, A. K., *The health of Glasgow, 1818–1925: an outline* (Glasgow 1930)

Chandler, T. J., *The Climate of London* (London 1965)

Clark, J., *The influence of climate in the prevention and cure of chronic diseases, mor particularly of the chest and digestive organs* (London 1830)

Clarke, E. (ed), *Modern methods in the history of medicine* (London 1971)

Clarke, R., *We all fall down: the prospect of biological and chemical warfare* (Londor 1968)

Clements, F. W. and Rogers, J. F., *Diet in health and disease* (Sydney 1966)
Clemow, F. G., *The geography of disease* (Cambridge 1903)
Clowes, W., *A short and profitable treatise touching the cure of the disease called morbus gallicus by unctions* (London 1579)
Cobbett, L., 'The decline of the death rate of diphtheria compared with that of scarlet fever', *Brit Med J*, 2 (1933), 139–40
Cobbett, W., *Rural rides* (London, 1830, 1853 ed)
Comrie, J. D., *History of Scottish medicine to 1860*, 2nd ed, 2 vols. The Wellcome Historical Medical Museum (London 1932)
Copeman, W. S. C., *Doctors and disease in Tudor times* (London 1960)
Court-Brown, W. M., Spiers, F. W., *et al*, 'Geographical variation in leukaemia mortality in relation to background radiation and other factors', *Brit Med J*, 1 (1960), 1753–9
Crawford, M. D., Gardner, M. J., and Morris, J. N., 'Mortality and hardness of local water supplies', *Lancet*, 1 (1968), 827–83
Crawfurd, R., 'Contributions from the history of medicine to the problem of the transmission of typhus', *Proc Roy Soc Med*, 6 (1913), 6–17
Creighton, C., *A history of epidemics in Britain, 1894*, 2 vols, 2nd ed, with additional material by D. E. C. Eversley, E. A. Underwood, and L. Ovenall (London 1965)
Currie, J., *Medical reports* (Liverpool 1797)
Darby, H. C. (ed), *An historical geography of England before 1800 AD* (Cambridge 1936, 1969)
Davidson, S. and Passmore, R., *Human nutrition and dietetics*. 3rd ed (London 1966)
Dickens, C., *Oliver Twist* (London 1838)
Defoe, D., *A journal of the plague year, 1722*. Introduction by D. J. Johnson (London 1966)
Dekker, T., *The plague pamphlets of Thomas Dekker*, ed F. P. Wilson (Oxford 1925)
Doll, R. and Hill, A. B., 'Lung cancer and other causes of death in relation to smoking', *Brit Med J*, 2 (1956), 1071
Doll, R. (ed), *Methods of geographical pathology*. Report of study group. International organisations of medical sciences. Blackwell Scientific Publications (Oxford 1959)
Drummond, J. C. and Wilbraham, A., *The Englishman's food. A history of five centuries of English diet*. Revised by D. Hollingsworth (London 1958)
Dublin, L. I., Lotka, A. J., and Spiegler, M., *Length of life: a study of the life table*. Rev ed (New York 1949)
Elkins, T. H., 'National characteristics of industrial landscapes', *Mélanges de Géographie physique, humaine, economique, appliquée, offerts à M. Omer Tulippe*, 11, ed J. A. Sporck and B. Schoumaker (Gembloux 1967)
Engels, F., *The condition of the working class in England in 1844* (London 1892)
Evelyn, J., *Fumifugium, or the inconvenience of the aer and smoke of London dissipated* (London 1661)
Ferguson, T., *The dawn of Scottish social welfare* (London 1948)
Ferguson, T., Benjamin, B., *et al*, *Public health and urban growth*, Report No 4. Centre for Urban Studies, University College (London 1964)

Fiennes, R. N. T., 'Stress in a crowded world', *New Society* (1963), 357
Finer, S. E., *The life and times of Sir Edwin Chadwick* (London 1952)
Fisher, F. J., 'Influenza and inflation in Tudor England', *Econ Hist Rev*, 2nd Ser, 18 (1965), 120–9
Fleure, H. J., *A natural history of man in Britain* (London 1951), rev ed 1971.
Forbes, D., 'Water-borne typhoid', *Lancet*, 1 (1938), 567–8
Fosberg, F. R. (ed), *Man's place in the island ecosystem. A symposium* (Honolulu 1963)
Fothergill, J., *An account of the sore throat attended with ulcers* (London 1748)
Fox, C., *The personality in Britain*, Nat Mus Wales (Cardiff 1947)
Frost, W. H. and Richardson, B. W., *Snow on cholera* (being a reprint of two papers) (London 1936)
Fussell, C. E., 'Agricultural and economic geography in the eighteenth century', *Geogr J*, 74 (1929), 170–7
Gale, A. H., *Epidemic diseases* (Harmondsworth 1959)
Gaskell, P., *The manufacturing population of England: its moral, social and physical conditions and the changes which have arisen from the use of the steam machinery* (London 1833)
Gilbert, E. W., 'Pioneer maps of health and disease in England', *Geogr J*, 124 (1958), 172–83
Glass, D. V., 'Some indicators of differences between urban and rural mortality in England and Wales and Scotland', *Public health and urban growth*, Centre for Urban Studies, University College, Report No 4 (London 1964)
Glass, D. V. and Eversley, D. E. C. (eds), *Population in history: essays in historical demography* (London 1965)
Goad, J., *Astro-meteorologica or Aphorisms and Discourses of the Bodies Celestial, their nature and influences . . . and other Secrets of Nature* (London 1686)
Graunt, J., *Natural and political observations . . . made upon the Bills of Mortality* (London 1662)
Greaves, J. P. and Hollingsworth, D. F., 'Trends in food consumption in the United Kingdom', *World Rev Nutr Diet*, 6 (1966), 34–89
Greenwood, M., *Epidemics and crowd diseases: an introduction to the study of epidemiology* (London 1935)
—— 'The epidemiology of influenza', *Brit Med J*, 2 (1918), 563–6
Gregory, S., 'Water supply maps for England and Wales', *Town Planning Rev*, 28 (1957)
Griffith, G. T., *Population problems of the age of Malthus* (Cambridge 1926)
Habakkuk, H. J., 'English population in the eighteenth century', *Econ Hist Rev*, 2nd Ser, 6 (1953), 117–33
Handschin, E., 'The effect of soil temperature on the behaviour and the migration of soil fauna', *Report on Agricultural Meteorological Conference, 1928* (London 1928)
Hare, R., *Pomp and pestilence: infectious disease, its origins and conquest* (London 1954)
Harris, W., *Tracatus de morbis acutis infantum* (London 1689). English translation by Cockburn, 1693
Harvey, G., *The disease of London or a new discovery of the scurvy* (London 1675)

Haviland, A., *Geographical distribution of disease in Great Britain*, 2nd ed (London 1892)
Hawkins, E., *Medical climatology of England and Wales* (London 1923)
Henschen, F., *The history and geography of diseases*; translated by J. Tate (New York 1967)
Hill, A. B. and Mitra, K., 'Enteric fever in milk-borne and water-borne epidemics', *Lancet*, 2 (1936), 589–94
Hirsch, A., *Handbook of geographical and historical pathology* (1881, in German), 3 vols. English translation by C. Creighton (London (Sydenham Society) 1883–6)
Hirst, L. F., *The conquest of plague: a study in the evolution of epidemiology* (Oxford 1953)
Hobson, W. and Pavanello, R., 'Air pollution in Europe', *New Scientist*, 268 (1962), 34–6
Hobson, W., *World health and history* (Bristol 1963)
Hodgkin, R. H., 'A history of the Anglo-Saxons', 3rd ed, 1 (1959)
Hort, E. C., 'Typhus fever', *Brit Med J*, 1 (1915), 673–5
Hoskins, W. G., *The making of the English landscape* (London 1963)
—— *Local history in England* (London 1959)
Howe, G. M., 'The geographical distribution of cancer mortality in Wales, 1947–53', *Trans and Papers, Inst Brit Geogrs*, 28 (1960), 190–210
—— 'Windchill, absolute humidity and the cold spell of Christmas, 1961', *Weather*, 17 (1962), 349–58
—— *A national atlas of disease mortality in the United Kingdom* (London 1963); 2nd rev and enl ed 1970
—— 'Computing the chilling effects of winter winds', *New Scientist*, 17 (1963), 276
—— 'Last month's cold snap', *New Scientist*, 28 (1965), 742
—— 'The geography of death', *New Scientist*, 38 (1968), 612–14
—— 'Geography looks at death', *Spectrum*, 71 (1970), 5–7
—— 'Disease patterns and trace elements', *Spectrum*, 77 (1970)
—— 'The mapping of disease in history', in E. Clarke (ed), *Modern methods in the history of medicine* (London 1971)
Huckstep, R. L., *Typhoid and other salmonella infections* (Edinburgh 1962)
Humphrys, G., 'Housing quality'. One of several maps published by the Population Studies Group, *Inst Brit Geogrs Trans*, 43 (1968)
Huxham, J., *Observations de Aere et Morbis Epidemica ab anno MDCCXXVIII ad Finem Anni MDCCXXXVII, Plymuthi factae his accedit opusculum de morbis colico damnoniensi auctore* (London 1739). (Translated from the Latin original: '*Observations on the air and epidemic diseases from 1728–37 inclusive made by Dr Huxham at Plymouth, together with a short dissertation on the Devonshire Colic* (Vol 1) 1738–48; (Vol 2) London 1759)
—— *An essay on fevers* (London 1750); 8th ed (Edinburgh 1779)
James, S. P., 'The disappearance of malaria from England', *Proc Roy Soc Med*, 23 (1929), 1–18
Jones, G. P., 'Cholera in Wales', *Nat Lib Wales J*, 10 (1957–8), 281–99
Kay, J. P., *The moral and physical condition of the working classes employed in the cotton manufacture in Manchester* (Manchester 1832)

Kopec, A. C., 'Blood groups in Great Britain', *Adv Sci*, 51 (1956), 200–3
—— *The distribution of the blood groups in the United Kingdom* (Oxford 1970)
Krause, J. T., 'Changes in English fertility and mortality, 1781–1850', *Econ Hist Rev*, 2nd Ser, 2 (1958–9), 52–70
Laidlaw, P., *Virus diseases and viruses* (London 1938)
Lamb, H. H., 'Our changing climate, past and present', *Weather*, 14 (1959), 299–318
—— *The English climate* (London 1964)
Langer, W. L., 'The Black Death', *Scientific American*, 210(2) (1964), 114–21
Lawther, P. J., 'Climate, air pollution and chronic bronchitis', *Proc Roy Soc Med*, 51 (1958), 262–4
Lawther, P. J., Martin, A. E., and Wilkins, E. T., *Epidemiology of air pollution: report on a symposium* (Public Health Papers, No 15) (Geneva 1962)
Lawton, R., 'Historical geography: the Industrial Revolution' (Ch 12), *The British Isles: a systematic geography*, ed J. W. Watson with J. B. Sissons (London 1964)
Lewis, W. P. D., 'Mortality from fog in London', *Brit Med J*, 1 (1956), 722
Longmate, N., *King cholera. The biography of a disease* (London 1966)
MacArthur, W. P., 'Some notes on old-time leprosy in England and Ireland', *J Roy Army Med Corps*, 45 (1925), 414–22
—— 'Old-time typhus in Britain', *Trans Roy Soc Trop Med Hyg*, 20 (1926–7), 487–503
—— 'Some medical references in Pepys', *J Roy Army Med Corps*, 50 (1928), 321–35
—— 'The identification of some pestilences recorded in the Irish Annals', *Irish Hist Studies*, 6 (1949), 172
—— 'A brief story of English malaria', *Brit Med Bull*, 8(1) (1951), 76–9
—— 'Medieval "leprosy" in the British Isles', *Leprosy Rev*, 24 (1953), 8–19
McKeown, T., 'Medical issues in historical demography', in E. Clarke (ed), *Modern methods in the history of medicine* (London 1971)
McKeown, T. and Brown, R. G., 'Medical evidence related to English population changes in the eighteenth century', *Population Studies*, 9(2) (1955), 119–41
McKeown, T. and Record, R. G., 'Reasons for the decline of mortality in England and Wales during the nineteenth century', *Population Studies*, 16 (1962), 94–122
McNalty, A., 'Indigenous malaria in England', *Nature* (17 April 1943), 440
Maitland, F. W., *Domesday book and beyond: three essays in the early history of England* (London 1960)
Malmesbury, Earl of, *Memoirs of an Ex-Minister* (London 1885)
Manley, G., *Climate and the British scene* (London 1952)
Marshall, H., 'Sketch of the new geographical distribution of diseases', *Edinb Med Surg J*, 38 (1832), 330–5
Maunder, W. J., *The value of the weather* (London 1970)
May, J. M., *Studies in medical geography. The ecology of disease.* Vol 1 (New York 1958)
—— *Studies in disease ecology* (New York 1961)
—— *The ecology of malnutrition in the Far and Near East* (New York 1961)

Mead, R., *A discourse on the plague* (London 1720)
Meetham, A. R., *et al*, *Atmospheric pollution: its origins and prevention*, 3rd rev ed (1964)
Medical Research Council, *Assay of Strontium 90 in human bone in the United Kingdom* (London 1969)
Meteorological Office, *Report on Agricultural Meteorological Conference, 1928* (London 1928)
Min Ag Fish, Food, 'Domestic food consumption and expenditure', *Rep Nat Food Surv Com* (1963–5)
Min Health, 'Report on the pandemic of influenza, 1918–19', *Rep Pub Health Med Subs* (1920), No 4
Morris, J. N., *Uses of epidemiology*, 2nd ed (London 1960)
Mourant, A. E., Kopéc, A. C., and Domaniewska-Sobczak, K., *The ABO blood groups* (Oxford 1958)
Mullet, C. F., *The bubonic plague and England: an essay in the history of preventive medicine* (Lexington, USA 1956)
Office of Health Economics, *The lives of our children: a study in childhood mortality* (London 1962)
—— *The price of poliomyelitis* (London 1963)
—— *Pneumonia in decline* (London 1963)
—— *New frontiers in health* (London 1964)
—— *The pattern of diabetes* (London 1964)
—— *Work lost through sickness* (London 1965)
Ooi, Jin-bee, 'Disease in a tropical environment', Ch 3 (42–77), *Rural development in tropical areas, with special reference to Malaya. J Trop Geogr*, 12 (1959)
Opie, I. and Opie, P., *Oxford dictionary of nursery rhymes* (Oxford 1955)
Osborne, R. H., Population (Ch. 18) in *The British Isles: a systematic geography*, ed J. W. Watson with J. B. Sissons (London 1964)
Parliamentary Papers (Commons), Hand-loom weavers: Assistant Commissioners' Reports. 1839 (159), 42, 511
—— (Lords), Report of the Poor Law Commissioners to the Secretary of State, on an Inquiry into the Sanitary Condition of the Labouring Population of Great Britain. 1842, xxvi, 1 (Reports by Rev W. Elkin and Dr Laurie)
—— (Commons), Reports by J. Smith and J. R. Martin in the Second Report of the Commissioners for inquiring into the state of large towns and populous districts. 1845 (602) (610), 18, 1, 299
Patrick, A., 'A consideration of the nature of the English sweating sickness', *Med Hist*, 9 (1965), 272–9
Pepys, S., *Pepys' Diary*, ed J. P. Kenyon (London 1963)
Percival, T., *Observations on the state of the population in Manchester* (Manchester 1773)
Petermann, A. H., *Cholera map of the British Isles, showing the districts attacked in 1831, 1832, and 1833.* Constructed from official documents (London 1852)
Pirie, N. W., 'Gluttony', *New Scientist*, 423 (1964), 838–41
Prothero, R. E., *The pioneers and progress of English farming* (London 1888)

Rawstron, E. M. and Coates, B. E., *Regional variations in Britain* (London 1971)
Reader's Digest Association, *Complete Atlas of the British Isles* (London 1965)
Redmayne, P., *Britain's food—the changing shape of things* (London 1963)
Rees, W., 'The Black Death in England and Wales, as exhibited in manorial documents', *Proc Roy Soc Med* (Hist Med), 16 (1923), 27
—— 'The Black Death in Wales', *Trans Roy Hist Soc*, 3 (1920), 115–36
Reiter, R., 'Neuere Untersuchungen zum Problem der Wetterabhängigkeit des Menschen', *Archiv fur Meteorologie, Geophysik und Bioklimatologie*, Ser B, 4 (Vienna 1953), 327–77
Renbourn, E. T., 'Normality in relation to human reactions', *Proc Ninth Int Congress on Industrial Medicine* (London 1948)
Richie, J., 'Enteric fever', *Brit Med J*, 2 (1937), 160–3
Roberts, R. S., 'A consideration of the nature of the English sweating sickness', *Med Hist*, 9 (1965), 385–9
—— 'The use of literary and documentary evidence in the history of medicine', *Modern methods in the history of medicine*, ed E. Clarke (London 1971)
Rodenwaldt, E., *Welt-Seuchen Atlas. Weltatlas der Seuchenverbrietung und Seuchenbewegung* ('World Atlas of epidemic diseases. World Atlas of the distribution and spread of epidemic diseases'), Parts I and II (Hamburg 1952, 1956)
Rogers, T. J. E., *The history of agriculture and prices in England.* Vol V (London 1866–7)
Rolleston, J. D., 'The history of scarlet fever', *Brit Med J*, 2 (1928), 926–9
Rowntree, S. B., *Poverty: a study of town life* (London 1901)
Russell, J. C., *British medieval population* (Albuquerque 1948)
Russell, W. T., 'The epidemiology of diphtheria during the last forty years', *Med Res Council, Special Reports Series*, 247 (1943)
Saltmarsh, J., 'Plague and economic decline in the later Middle Ages', *Camb Hist J*, 7 (1941), 23–41
Schroeder, H. A., 'Relation between mortality from cardiovascular disease and treated water supplies', *J Amer Med Ass*, 172(17) (1960), 1902–8
Scott, H. H., *Some notable epidemics* (London 1934)
Shapter, T., *The history of the cholera in Exeter in 1832* (London 1841)
—— *Sanitary measures and their results: being a sequel to the 'History of cholera in Exeter'* (London 1853)
Shaw, M. B., 'A short history of the sweating sickness', *Ann Med Hist*, 5 (1933), 246–73
Shrewsbury, J. F. D., 'The yellow plague', *J Hist Med Allied Sci*, 4 (1949), 5–47
—— 'Henry VIII: A medical study', *J Hist Med Allied Sci*, 7(2) (1952), 141–85
—— *A history of bubonic plague in the British Isles* (London 1970)
Siegfried, A., *Germs and ideas: routes of epidemics and ideologies.* Translated by J. Henderson and M. Claraso (Edinburgh and London 1965)
Singer, C., *A short history of medicine* (Oxford 1928)
Smith, W. D. L., 'Malaria and the Thames', *Lancet*, 270 (1956), 433–6
Snow, J., *On the mode of communication of cholera* (London 1849); 2nd ed (1855)

Spiers, F. W., *The hazards to man of nuclear and allied radiations*. Report by a committee appointed by the Med Res Council (London 1956)

—— *The hazards to man of nuclear and allied radiations*, II. A second report to the Med Res Council (London 1960)

Stamp, L. D., *The geography of life and death* (London 1964)

Stocks, P., *Regional and local differences in cancer death rates* (Studies in Med and Pop Subjects) (London 1947)

Talbot, C. H., *Medicine in medieval England* (London 1967)

Tanner, J. W., 'Earlier maturation in man', *Scientific American*, 218 (1968), 21–7

Thackrah, C. T., *The effects of arts, trades, and professions and of civic states and habits of living on health and longevity* (London 1831)

Thompson, J. W., 'The aftermath of the Black Death and the aftermath of the Great War', *Amer J Sociol*, 16 (1920–1)

Thomson, G., *Loimotamia, or the pest anatomized* (London 1666)

Thomson, J. and Abbott, D. C., 'Pesticide residues: history, alternatives and analysis', *Roy Inst Chem*, Lecture Series No 3, 1966 (London 1967)

Thomson, D., 'Whooping cough: a review', *Monthly Bull Min Health*, 12 (1953), 92–102

Thring, M. W. (ed), *Air pollution*. Based on papers given at a conference in Univ of Sheffield, 1956 (London 1957)

Trevelyan, G. M., *English social history: a survey of six centuries, Chaucer to Queen Victoria*. 3rd ed (1961)

Tromp, S. W., *Medical biometeorology* (London 1963)

Tromp, S. W. and Weihe, W. H. (eds), *Biometeorology* (London 1967)

Tucker, G. S. L., 'English pre-industrial population trends', *Econ Hist Rev*, 2nd Ser, 16 (1963), 205–18

Tyler, W. F., 'Bracing and relaxing climates', *Q J Roy Met Soc*, 61 (1935), 309–15

Underwood, E. A., 'The history of cholera in Great Britain', *Proc Roy Soc Med* (1948), 165–73

United Nations, *Demographic year book, 1967* (New York 1968)

Vogel, F., 'ABO blood groups and disease', *Amer J Human Genet*, 22 (1970), 464–75

Waddy, B. B., 'Climate and respiratory infections', *Lancet*, 2 (1952), 674–7

Warren, H. V., 'Geology and health', *Sci Monthly*, 78(6) (1954), 339–45

—— 'Geology and multiple sclerosis', *Nature*, 184 (1959), 561

Warren, H. V., Delavault, R. E., and Cross, C. H., 'Possible correlations between geology and some disease patterns', *Ann New York Acad Sci*, 136 (1967), Art 22, 657–710

Watson, J. W. with Sissons, J. B. (eds), *The British Isles: a systematic geography* (London and Edinburgh 1964)

Watt, R., *Treatise on the history, nature and treatment of chin-cough* (Glasgow 1813)

Weiner, A. S., 'Blood groups and disease', *Amer J Human Genet*, 22 (1970), 476–83

Wells, C., *Bones, bodies and disease* (London 1964)

Willis, T., *Diatribe dual* (The Hague 1659)

Wilson, F. P., *The plague in Shakespeare's London* (Oxford 1927)

Winslow, C. E. A. and Herrington, L. P., *Temperature and human life* (London 1949)
Wise, M. E., 'Human radiation hazards', *Physicomathematical aspects of biology*, ed N. Rashevsky (New York 1962)
Woodham-Smith, C., *Florence Nightingale, 1820–1910* (London 1950)
Woods, H. M., 'Epidemiological study of scarlet fever in England and Wales since 1900', *Med Res Council, Special Reports Series*, 180 (1933)
World Health Organisation, 'Endemic goitre' (Geneva 1960); Basic documents. 16th ed (Geneva 1965)
Yudkin, J. and McKenzie, J. C. (eds), *Changing food habits* (London 1964)
Ziegler, P., *The Black Death* (London 1968)
Zinsser, H., *Rats, lice and history* (London 1935)

Index

References to maps, graphs and tables are distinguished by italic type. When the subject is mentioned in **Notes and References,** the letter *n* is added to the page number.